GROWING UP

Some other books by the same author

Pastoral Properties of Australia
Thoroughbred Studs of Australia and New Zealand
Australia: The First Twelve Years
An End to Silence: The Building of the Overland Telegraph Line
An Australian Country Life
Springfield: The Story of a Sheep Station
Food From Far Away
A Taste of Australia
A Celebration of Shore
Station Life in Australia: Pioneers and Pastoralists
The Atlas of Australian History

For children
How People Lived
How People with Cattle Settled the Outback
Keeping in Touch with Each Other
On the Sheep's Back
The Bush Pioneers
The Bushrangers
The Convict Settlement
The Goldrush Era

GROWING UP

Forestry in Queensland

Peter Taylor

ALLEN & UNWIN

© Department of Primary Industries Forest Service, Brisbane 1994

This book is copyright under the Berne Convention.
No reproduction without permission. All rights reserved.

First published in 1994

Allen & Unwin Pty Ltd
9 Atchison Street, St Leonards, NSW 2065 Australia

National Library of Australia
Cataloguing-in-Publication entry:

Taylor, Peter, 1936– .
 Growing up: forestry in Queensland.

 Bibliography.
 Includes index.
 ISBN 1 86373 751 0.

 1. Queensland Forest Service. 2. Forests and forestry – Queensland.
 I. Title.

634.909943

Set in 11/12.5pt Baskerville by DOCUPRO, Sydney
Printed by Australian Print Group, Maryborough, Victoria

10 9 8 7 6 5 4 3 2 1

Contents

Introduction	viii

PART I
1	The Ever Changing Forest	3
2	Who Needs Trees?	17
3	The Limitless Resource	32
4	Forever is a Long Time	45
5	The Swain Era	58
6	The War and the Aftermath	73
	Profile: Victor Fedorniak—Cossack–Australian	85
7	Changing Needs, Changing Views	89
	Profile: Mark Peacock—Forester	102

PART II
8	Managing Native Forests	109
	Profile: Sam Dansie—Retired Forester	122
9	Plantations	126
	Profile: Training the Foresters	138
10	Fire	142
11	Fraser Island	157
	Profile: Aila Keto—Conservationist	170
12	Sawmilling	174
	Profile: John Crooke—Sawmiller	187
13	Changing Uses	191

	Profile: Andy McNaught—Scientist	203
14	Technology and the Forest	207
15	The Future	220

Appendix One
    Heads of the Queensland Forestry Department   226
Appendix Two
    Definitions of Some Forestry Terms Used in the
    Conservation Debate   227
Appendix Three
    Organisations Associated with Forestry   230

Bibliography   234
Index   238

I happily dedicate this book to Peter Holzworth,
forester, poet and friend

*Unless otherwise acknowledged, all the photographs in this book
are from the library of the Queensland Department of
Primary Industries Forest Service*

Introduction

Anybody driving along the Bruce Highway in southern Queensland will be aware of forests. The road is lined with them, often on both sides, and they keep you company for many kilometres. Most are plantations, and most of the trees are pines.

This book sets out to explain forestry in Queensland: how and why it started, how it developed and what it does now. This explanation is directed to the general reader, curious perhaps about all those trees along the highway. It is not aimed at those involved with forestry, although they may find some of it useful.

It might seem that growing trees is one of the simplest things to do. It probably is. But growing forests is not. Forestry is a many faceted activity which now involves technology across many disciplines and skills of many kinds. In these respects forestry is far from simple, and it becomes less so almost every day. Although this is not a technical book, the subject it deals with is technical. This is where the distinction between the general reader and the professional forester is most acute. I have tried to explain some of the technicalities for the general reader and I hope I have not offended the truth in the process.

Foresters have a very large body of technical literature of their own, but there have been few attempts to explain forestry to a wider audience. Perhaps this book will go some way towards doing that.

I have taken two liberties in the text which require explanation.

INTRODUCTION

I have sometimes used the word forester to mean anybody engaged in forest work, which is a wider use than it bears in forestry. A forester is a person with a Bachelor of Science degree in forestry, but I could find no ready alternative when I wished to talk collectively of those involved in forestry.

The other is that I have referred throughout the book to the forestry department. It carried many other titles during the last ninety years. In recent years it was known as the Queensland Forest Service and it is often still referred to as such by older people. Reflecting the changes in name as they occurred would have been confusing, so I use forestry department for simplicity and consistency.

The use of metrics always poses a problem when dealing with historical material. Where money was expressed in pounds I have let it stand as converting to dollars is meaningless. Where possible I have converted measurements to their metric equivalent but some, such as super feet per acre, I have left alone. The conversion to cubic metres per hectare is clumsy and the original measurements used in this book are usually used to show changes in one direction or another. These changes can be seen just as easily in the old measurements, which were usually rounded approximations anyway. Measurements in quoted material have been retained in their original form.

I am not a forester, nor am I a conservationist. But in the process of writing this book I have come to respect both.

I have also come to love the forests. I thought I always did, in common with most Australians. But I have had the privilege of seeing forests through the eyes of foresters, and through them developed an awareness that I did not have before.

This book could not have been written without the help and support of the Forest Service in Queensland. However, it is fair to say that the views expressed in this book are, of course, mine alone. The staff of the Forest Service were unstinting in their help and enthusiasm. There are too many to list and so I offer my thanks to them collectively.

I also thank Mary Dettman of the University of Queensland for her help with Chapter One, and John Perlin for allowing me to draw on his book *A Forest Journey* in Chapter Two.

I received much help from the timber industry and again I thank all those people who freely gave their time and who made sure I did not wander too close to their saws.

Finally, and by no means least, I thank my wife Rosemary. While I was working on this book she learned far more about forestry than she thought she needed to know. Her comments on the early

GROWING UP

manuscript did much to shape the finished book and I am grateful for her participation.

Peter Taylor
Runaway Bay
1994

PART I

A fossil of a kauri pine. This specimen is over 170 million years old but the features are similar to the present-day kauri.

1 The Ever Changing Forest

A forest is awe-inspiring, especially if it consists of tall timber and is seen on a bright sunny day. It seems to demand some emotional response from those who venture into it and the response is usually forthcoming. Most people are moved and impressed by what they see and the reaction can be spiritual, aesthetic and uplifting. Very few remain indifferent.

Although these responses are intensely personal, and therefore almost infinitely varied, there is a thread that links most of them. There is usually a sense of timelessness, a sense that one is looking at something that has remained unchanged from time immemorial and which will still be there when we are long dead.

While this is true in human terms, it leads to a perception that forests are timeless, unchanging and static: that what we see now has always been so and that it will remain so for eternity.

There is another perception that is closely related to the first, and that is that before the arrival of Europeans Australia was almost entirely covered with forest. The fact that forests now occupy only a small proportion of the landmass is assumed to be a result of European destruction.

Neither of these perceptions is true, although human activity in Australia has certainly destroyed vast areas of forest.

The truth is that forests are not static. They have not always been as we see them and it is unlikely that they will remain for ever as they are now. Nor was Australia a heavily forested land before Europeans arrived. Surprising as it may seem, there were probably more trees in Australia twenty years after the arrival of Europeans than there were when they came.

Because human beings are able to dominate their environment, or think they can, we have developed an arrogance that comes from superiority. And because of that arrogance we have difficulty imagining a world without human beings, and the changes that world went through before we appeared. The magnitude of those changes, and the huge time scale that they occupied, challenge our feeling of superiority. If changes of that order could take place without us, then perhaps our superiority is not as well based as we might like to think.

The Earth is about 4600 million years old. But numbers such as that are difficult to comprehend. How big is 4600 million? An astronomer deals in such numbers, as do other specialists, but most of us do not. They are beyond our comprehension. So a little imagery might be useful.

Let's assume that the history of the world is shown on the face of a clock, and that the Earth was created at one minute past twelve. It is not until nearly 3 o'clock that the first fossils appeared. From then until just after 9 o'clock the Earth was inhabited by primitive life forms which collectively produced an atmosphere on which more sophisticated life forms later depended. These sophisticated life forms had developed by about 9.15 but they were confined to the sea. It was not until nearly 11 o'clock that life moved on to the land. Finally, at a few seconds before 12 o'clock, human beings appeared. This was about four million years ago. *Homo Sapiens* appeared only within the last million years and arrived in Australia about 40 000 years ago.

The influence humans had on the Earth is limited to a very small part of the time scale. The Earth was subjected to forces far greater than humans and had already changed significantly before they arrived.

It is only recently that some of these changes have been identified with any certainty. Geologists and botanists recognised that there were similarities between parts of the world that are now widely separated. These similarities were so strong as to suggest that these parts might once have been joined together and that they had somehow drifted apart in more recent times.

The difficulty was in explaining how such movements could have occurred. The concept that huge masses of land could somehow have moved across the face of the Earth was difficult to accept, no matter how convincing the evidence might be.

This was not overcome until technology led to the theory of plate tectonics. This suggests that the Earth's crust is made up of about fifteen separate plates which have not only moved in the past but which are still moving now.

Originally these plates were joined in a supercontinent called Pangaea. This continent provided a common mass for all land-based life forms and it was here that they competed for their biological survival. About 200 million years ago Pangaea started to break up and after about 20 million years it had split into two separate continents: Laurasia in the north and Gondwana in the south. Not long afterwards Gondwana also started to break up. India separated from this landmass and started moving across the sea until it joined the Asian mainland. The collision of these plates formed the Himalayas.

Other landmasses also broke away from Gondwana. One consisted of South America and Africa and another of Antarctica and Australia. As these drifted further apart newly evolved life forms were confined to one or the other.

About 49 million years ago Australia broke away from Antarctica and slowly drifted north to arrive at its present location. New Zealand was rotated and the edge of its plate pushed beneath eastern Australia to form the Great Dividing Range. Eventually New Zealand broke free, but it was not until about 8000 years ago, when the ice was melting, that Tasmania and New Guinea finally separated from the Australian mainland.

During this time biological development was also taking place. Every living form is derived from another and not from a separate spontaneous creation. They are therefore linked in a chain of reproduction. Some life forms survived, some did not. Most of those that survived (but not all) did so because they were able to adapt to the changing environment.

These changes were slow but dramatic. Lakes and swamps covered much of Australia while it was still part of Gondwana. The Great Australian Basin, for example, covered over 770 000 square kilometres of the interior and lasted until about 100 million years ago, when much of Australia was inundated by the sea. Although this inundation was relatively shallow it significantly reduced the living space for land-based life forms. Later, when Australia became a separate landmass, the survivors of this flood developed those characteristics which we now recognise as being uniquely Australian.

When Australia separated from Antarctica most of its vegetation consisted of cool-temperate leathery leaved rainforest which is now known as the Gondwana Element. As Australia drifted further from

Antarctica warm-temperate and tropical forest spread across southern and eastern Australia, but about 30 million years ago the temperature dropped sharply and the tropical components disappeared from the rainforests in the south and other species took their place. The tropical rainforests now confined to the north-east coast include the survivors of the late Cretaceous forest.

Meanwhile the landmass was being shaped by erosion. This eventually reduced much of the land to a featureless plain which was so flat that there was no drainage and no rivers. It was already an old and worn country that was unlikely ever to change. But it did change. In the Pleistocene period, which began about two million years ago, rainforest persisted in some areas of the centre. The climate was cool, there was reliable rain, rivers ran and life flourished. But between 20 000 and 10 000 years ago there was an arid phase and things changed again. The climate became warmer, rainfall was reduced, rivers ran dry and lakes dried out to leave small areas of salty water that could not support life.

As these conditions spread the dry-living plants (Xerophytes) began to dominate Australia. These plants had thick or needle like leaves, or thick bark containing resin to protect the sap. They had large root systems and their fruits were hard and produced seeds which retained their fertility in arid ground until rain arrived to germinate them.

Xerophytes belong to three orders. These are Proteaceae, which include banksias, hakeas and waratahs; Casuarinaceae, which are unique to Australia; and Myrtaceae, which include all tea-trees and eucalypts.

Over 600 species of eucalypts grow in Australia and they probably spread from the northern tropics to the temperate south. While not unique to the continent they are so universal as to be an essential element in any Australian landscape. Although different individually, eucalypts all shed leaves and bark continuously and are very resistant to insects, fires and humans.

During the last 750 000 years Australia, and much of the rest of the world, has experienced eight major climatic changes, from glacial to interglacial, and these changes have had a profound effect on the distribution and population of trees.

At the height of the last interglacial, which was between 25 000 and 10 000 years ago, the cold and aridity reduced tree population to about 15 per cent of the landmass. Forests retreated down the mountainsides and the red gum forests disappeared from southern streams. Some small areas changed less than the rest. Pockets of trees

THE EVER CHANGING FOREST

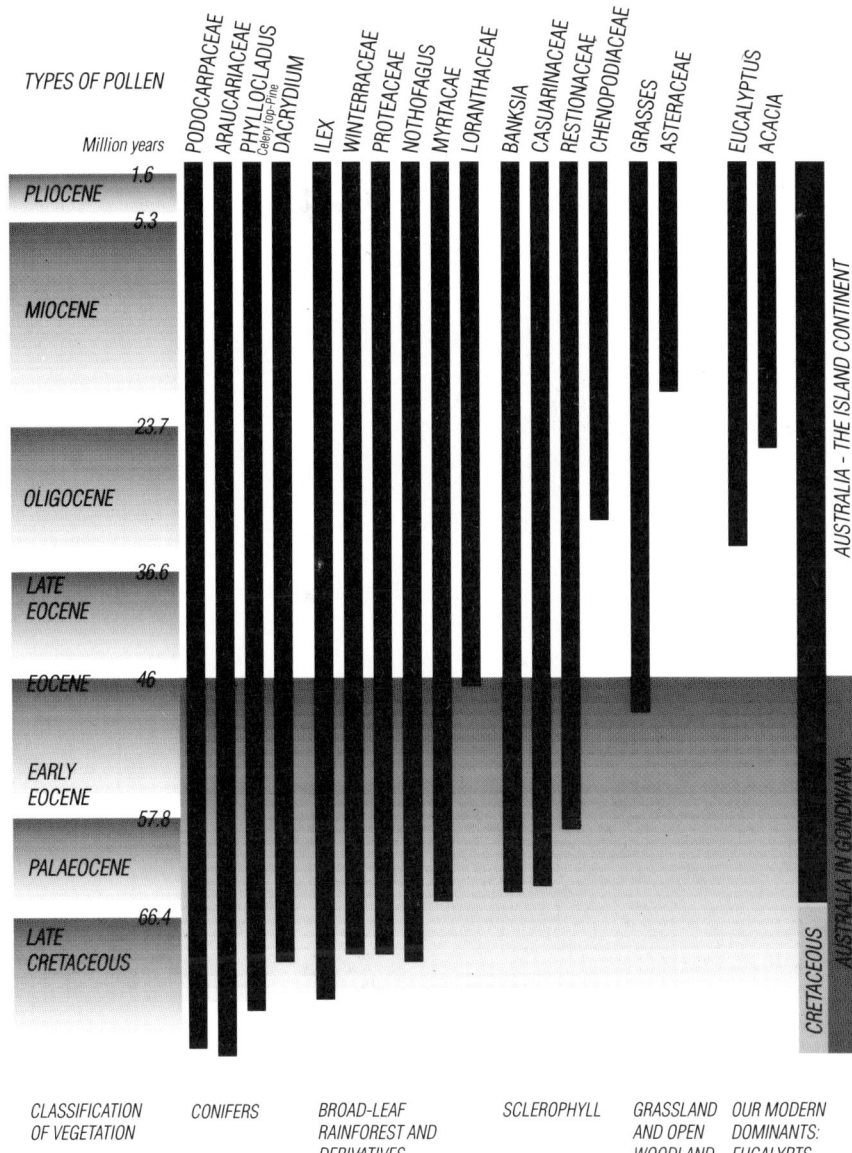

This chart shows the age of fossilised pollens. Note that some pollens do not produce fossils and therefore do not appear in a chart of this type.

survived in their original locations and many of their descendants remain to confuse us.

During the last 10 000 years forests changed yet again and became more extensive as they advanced onto new ground. Trees

7

marched back up the hillsides, sometimes leaving snow gums standing in lower valleys. Arid and sandy coastal areas were occupied by casuarina, acacia and mallee on a big scale and, again, remnants were left behind in eucalypt forests.

In the tropics, warmer and wetter conditions led to a recovery of the rainforests, which invaded areas previously occupied by other types of forest. This was particularly so between 7000 and 4000 years ago, which suggests that conditions were wetter than they are now. Certainly many patches of forest have declined in area in recent times but moist forests are still more extensive than they were in some previous ages. Interglacial periods, such as we are in now, produce much more expansion of lowland tropical forests than do glacial periods.

These changes had nothing to do with human beings. Many happened before the arrival of humans and those that happened afterwards were on a scale that was completely beyond the ability of humans to alter. They were also slow. The lifespan of a single human being would notice little difference, but they were taking place just the same, as they still are. This is not to suggest that human beings had no effect. They obviously did. But they had no effect on these major changes that were slowly going on around them.

The change that humans did bring about was in the use of the land. During the last 10 000 years agriculture developed in one form or another across most of the world. The discovery that crops could be raised and animals reared was one of the great advances of humanity. This meant that people no longer had to live as nomads. They could produce food instead of having to find it. Settlements could be formed. Indeed they had to be, because tending crops and livestock tied people to their demands and needs. Where agriculture was successful it marked the end of the nomad and the start of the human effect on the earth. This was insignificant compared to the global climatic changes that were also taking place, but it was there nevertheless.

Except in Australia.

The Australian Aborigines were unique in that they did not adopt agriculture in the usual sense of the word and they remained as food-gathering nomads until the arrival of Europeans, and in many cases long after.

Aborigines probably have the longest continuous cultural history in the world, and that history has always been tied to the land. Although Aboriginal culture was extremely complex and varied from one part of the country to another, there were some elements in common. One was the structure of Aboriginal societies, and particularly the association with a specific tribe. Within each tribe were smaller social groups based on families or clans.

At the time of European settlement it is thought that there were about 600 tribes in Australia and that they were spread over the entire continent. Each tribe occupied a clearly defined area which was of great significance to all its members, and this area was recognised by neighbouring tribes.

The relationship between the tribe and the ground it occupied was of fundamental importance. Aborigines knew that the world had been made by their ancestors during the Dreamtime, and that when their ancestors had finished they became embodied in the landscape in the form of hills, rocks, creeks and other natural features. The difficulties the ancestors had encountered also had precise locations and were part of the topography.

The stories of this creation were well known to all members of the tribe, although some were not revealed to women and children. These stories were part of the being of all Aborigines, as was the physical presence of their ancestors. When Aborigines looked at a landscape they saw tangible signs of their ancestors. They related to them and knew their achievements instinctively, not in a historical sense but as part of their daily lives.

It is obviously difficult to estimate the Aboriginal population at the time of European settlement. Figures offered range from about 100 000 to 900 000. What is known more certainly is that population densities varied considerably. The interior of South and Western Australia was only lightly occupied, with one suggestion that only 18 000 people occupied an area of 650 000 square kilometres. Elsewhere, in coastal and river areas and parts of the north, it has been suggested that densities might have been as high as three people to a square kilometre.

Aborigines were never farmers in the Western sense of the word. They did not grow crops or raise livestock. Instead they depended entirely on whatever nature had to offer and they covered large tracts of land in search of it. They certainly understood the effect of seasons on the food they sought and knew exactly when and where certain seeds, fruits and insects would be available. So they took advantage of natural processes without intervening in them. There was, however, one exception. They used fire, and they used it for thousands of years.

Burning the land produced lush growth of young grass, which in turn attracted kangaroos and other game that formed an important part of their hunting. It also encouraged certain seeds to germinate which produced edible roots or fruits, or which might be

GROWING UP

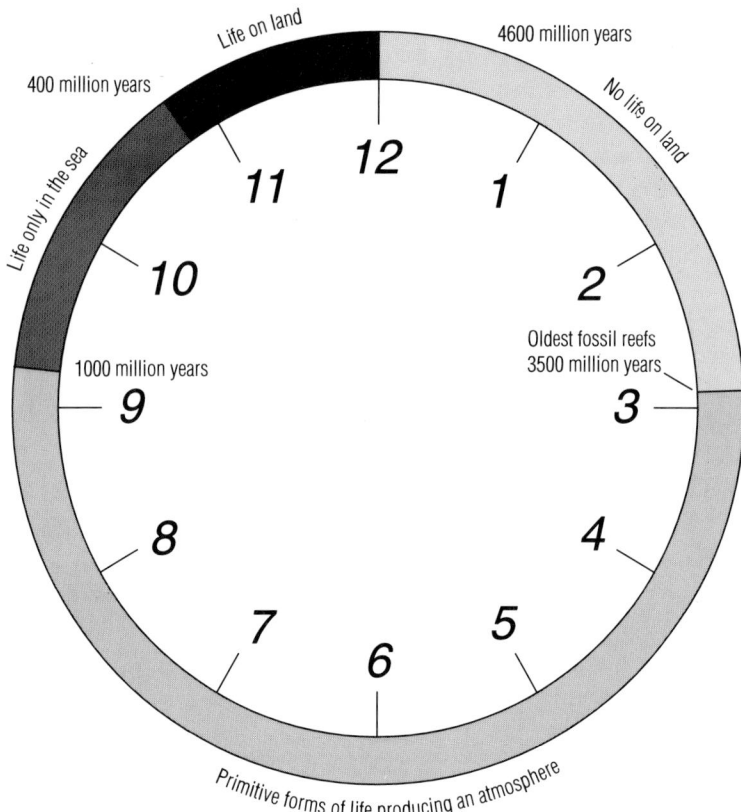

The development of the Earth shown on a clock face from 12 midnight to 12 noon. On this scale, humans appear only a few seconds before noon.

attractive to insects which were themselves edible. Burning also kept the land open, so that while there was still enough cover for game, it did not become so dense as to make hunting impossible. Burning could also be used to drive game towards waiting hunters.

Before the arrival of Europeans, burning by Aborigines was universal and constant although not haphazard. Practically every European explorer commented on the use of fire by Aborigines. Indeed, smoke from a burning landscape was often the first indication of the presence of Aborigines. Abel Tasman saw Aboriginal fires, as did Cook, who reported fires all the way along the east coast and commented on the ability of the Aborigines to fire large expanses of country. Oxley saw burning during his journeys in northern and central New South Wales in 1820; Stokes saw it in the south-west of Western Australia in 1846 and Leichhardt saw it in south-east Queensland at about the same time.

In 1848 Thomas Mitchell described in his journal the significance of fire to the Aborigine:

THE EVER CHANGING FOREST

> Fire, grass, kangaroos, and human inhabitants, seem all to be dependent on each other for existence in Australia; for any one of these being wanting, the others could no longer continue. Fire is necessary to burn the grass, and form the open forests, in which we find the large forest-kangaroo . . . the omission of the annual periodic burning of the grass and young saplings, has already produced in the open forests near Sydney thick forests of young trees, where, formerly, a man might gallop without impediment, and see whole miles before him. Kangaroos are no longer to be seen there; the grass is choked by underwood; neither are there natives to burn grass.

That description is important for two reasons. The first part shows the effect that burning had on the country, and the second part describes what happened when it was no longer burnt.

By burning the bush the Aborigines changed the landscape. The trees which were fire resistant survived, and so eucalypt forests were widespread. Other species such as pines and casuarinas are very sensitive to fire and did not survive in areas that were regularly burnt. As these disappeared, their space was occupied by the eucalypts.

Although these changes were relatively insignificant compared to the climatic changes that were also taking place, over many years of settlement in Australia the Aborigines changed the vegetation and the appearance of much of the country. Their fires reduced the size and variety of the forests and exposed the soil which in turn led to erosion.

The country the Aborigines found was originally more diverse. Pollen evidence indicates that there was a much wider distribution of plant growth than there was at the time of European settlement because fire sensitive species had not been driven out.

By burning the land, Aborigines made it more productive. As a result of burning the country was better able to produce the food they needed. In changing the nature of the land, they made it more useful. This hardly makes them agriculturalists because it was the only method they used, but their intentions had much in common with more developed forms. And, like agriculture elsewhere, the land was changed. So much so that much technical effort is required to determine what it was like before this regime started. There is, however, no shortage of descriptions of what it was like when Europeans arrived.

Many early settlers in the south-east described the land as park-like. This was a European term describing lightly timbered country with an abundance of grass and occasional shrubs. It was, of course, very attractive to those of European origin not only because of its pleasant and familiar appearance but also for what it promised in terms of agricultural production. Had European settlement taken place 20 000 years earlier the land would probably have looked far less attractive to European eyes because the change would not have reached that stage. The change that Aborigines produced on the land had the unwelcome effect of making it attractive to Europeans who, not surprisingly, took possession of it.

The second part of the description in Mitchell's journal describes what happened when the land was no longer burnt. Burning land was not an agricultural technique used by Europeans except, occasionally, to clear it and so the land they settled was no longer burnt. The result was uncontrolled growth that was too prolific to be contained. Forests became thicker and more extensive, undergrowth turned rank and game disappeared. So the forests that had been changed by generations of burning by Aborigines reverted to a stage they had passed through on the way. Not exactly so, because although fire sensitive species might in time have re-established themselves, the activity of Europeans started to change the landscape yet again.

The emotional human response to forests is perhaps never sharper than when the forest is a rainforest. Rainforest has a quality that is usually lacking in most other types. The trees are often very tall and the cover they form at that considerable height is almost total. Because of this not much light penetrates to the forest floor and ground vegetation is sparse. All this leads to an almost religious response. People talk of being in a natural cathedral and it is a fair comparison. The forest itself is quiet and invites the same reverence.

This surely is the timeless, unchanging forest which has stood here since the beginning of time. But, magnificent though it is and threatened though it might be, rainforest is neither timeless nor unchanging.

As we have seen, at various times in the past rainforest occupied a much greater area of Australia than it does now, and at other times it occupied less. This movement is not yet fully understood. About 35 million years ago rainforest was widespread but by no means universal. It seems to have contracted slowly, because of climatic changes, and continued to do so at a more rapid rate until 200 000 years ago.

These climatic changes included lower temperatures, and this was true of the world as a whole. These changes were not constant and they varied in different parts of the world. But between fifteen million and five million years ago there was a significant fall in temperature, and between two million and one million years ago this developed a kind of rhythmic change in which long cold periods were interrupted by short warmer ones, one of which we live in now.

THE EVER CHANGING FOREST

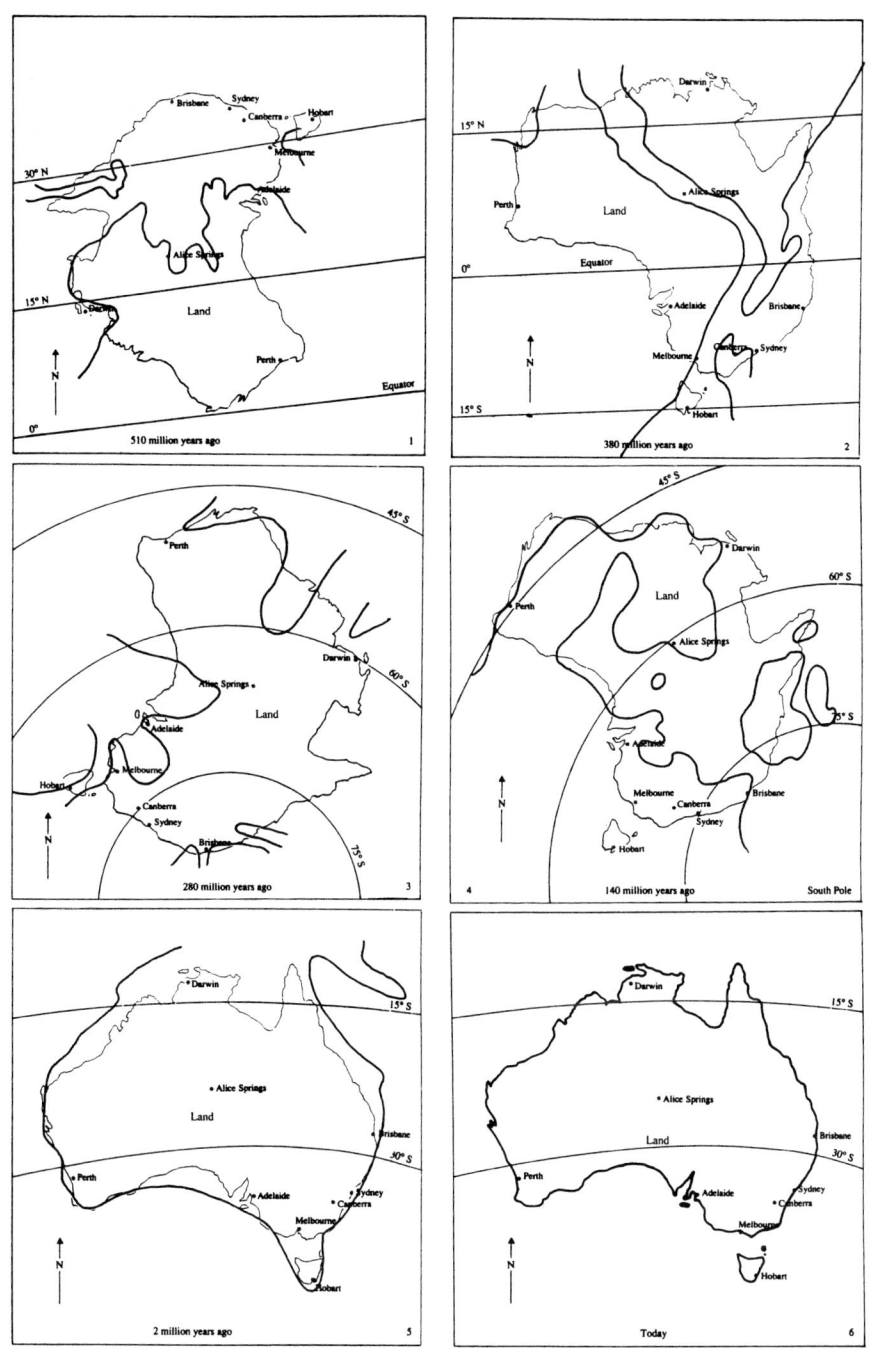

These maps show the position of Australia at various times in the past. The thick lines show the area of land at each stage.

During these changes the weather also changed. Rainfall varied both in amount and location, and the height of the seas also changed.

13

GROWING UP

In northern Australia the last cold period occurred about 18 000 years ago. At that time the shoreline of eastern Australia was approximately the outer edge of the Great Barrier Reef. New Guinea was connected to Australia by land and the Gulf of Carpentaria was a landlocked lake. It is only about 6000 years ago that the sea rose to its present height.

Not surprisingly, the rainforests were affected by these considerable changes, slow though they were. But establishing exactly how they were affected is no easy matter. Most information comes from pollen samples but dating them is difficult if they are older than about 40 000 years, which is the limit of radiocarbon dating.

Lynch Crater, on the Atherton Tableland, has been intensively studied to reconstruct the changes that occurred there. Pollen records spanning the past 200 000 years indicate that 27 000 years ago some type of rainforest grew around the crater. But it was not always the same type. Pollen evidence shows that some species were constant throughout this period, while others are from species which are now extinct. So although there was rainforest during that time, it was constantly changing.

Between 27 000 and 7000 years ago the rainforest disappeared and it was replaced by a eucalypt savanna. Parts of the rainforest began to return and it took between 1000 and 1500 years to become fully established again on each site. So the rainforest that the first Europeans saw there had been growing for only about 7000 years, even though rainforest had previously occupied the site nearly 200 000 years earlier and for a considerable time. So we cannot assume that an area which is now occupied by rainforest has been continuously occupied by similar rainforest 'for ever'. Nor can we assume that the rainforested area of northern Queensland was always bigger than it is now. The extent of rainforest has clearly changed and will continue to change.

There are, of course, also changes within the rainforest itself. A cyclone for instance will defoliate the canopy and blow down a number of tall trees. This allows light to enter part, or all, of the forest and certainly to places where there was no sunlight before. This light germinates long-dormant seeds of pioneer species which grow rapidly and make that part of the forest look very different to what it was before. The shade they provide then permits the germination of rainforest species. These species, being dominant by nature, overtake the pioneers after a few years and compete amongst each other for the light. The pioneers cannot survive beneath this new canopy and die out.

This process might take no more than a decade but while it is

THE EVER CHANGING FOREST

going on the forest will look different. A trained observer will see very clearly what is happening but a less informed observer might think that the rainforest is being destroyed by other species. After about twenty-five years the process of repair will probably be complete and even the trained observer might see no evidence of the original damage, or even know that there had been any.

The history of Aborigines in Queensland rainforests is largely unknown. Archaeological studies are relatively recent and so far the oldest evidence is from about 5000 years ago. However, because of the changes to the rainforest area there might well have been much earlier occupation of areas that were forested then but which are not now. Far more research needs to be carried out before a more complete picture can emerge.

What does seem certain is that the Aboriginal settlement of North Queensland rainforests was quite dense, which in turn suggests abundant resources. Fish and game seem to have existed in profusion, as did vegetable plants. Some of these plants were very toxic and although they made up the staple diet they could be used only after suitable treatment. Aborigines in other parts of Australia and in the rest of the world knew how to treat toxic plants to make them edible, but the range of such plants in northern Queensland was far greater than anywhere else.

It is unlikely that the Aborigines used fire in the rainforests. First, there would be little need. The forest floor remained fairly clear and the plants that did grow there were capable of reproducing without the need for encouragement. Second, rainforests are not very flammable and it would be quite difficult to put a fire through one, but if they do burn most species have no resistance and die as a result.

So, dramatic as most Australian forests are, they are not timeless and neither did they cover the whole of the continent before the arrival of Europeans.

On the contrary, the forests have always been in a state of change as a result of climatic changes. They have expanded and contracted and sometimes disappeared completely. The species making up the forests have also changed from time to time as conditions have favoured some at the expense of others.

Far from being timeless, the forests have never stopped changing and presumably never will. The fact that these changes take place on a timescale much greater than can readily be comprehended by human beings means that they are easily overlooked. During a single lifetime the forest might show no sign of change apart from the unlucky effects of wildfire or storm, and even these will be repaired fairly quickly.

They are changing constantly as a result of other natural changes, but the greatest source of change is humans. Aborigines changed the forest by their periodic burning. Indeed they changed the whole landscape. But that was little compared to the changes brought about by the arrival of Europeans and which is described in later chapters.

Humans were the major threat to forests. Forests could cope with natural changes but they had no defence against human beings. It was only when people realised that the forests might disappear in their own lifetimes that attitudes began to change.

In recent times we have become aware of the need to conserve forests. But even that is not as simple as it might sound. In what state are they to be conserved? As they are now? As they were before the arrival of humans? Even if we had the means of preserving them in amber in any state of our choice, we might ignore the fact that they are always in a process of change. Forests are living, dynamic beings and must be allowed to remain so.

Meanwhile, let us hope that we can continue to enjoy them and the emotions that they create in us. They are our contact with times past and a reminder of our brief span.

2 Who Needs Trees?

Who needs trees? We do. We always have and probably always will.

Trees were our original resource. They were there when human beings first appeared on earth and they have been used constantly ever since.

Those early humans were limited to what resources they could see or what experience showed was there even though it could not be seen. There were not many of these. Edible roots could not be seen, but the plants they supported were an infallible indication of their presence. Fish could rarely be seen directly, but their presence was visible by swirls in the water or the presence of certain birds. So what these early human beings saw was what they had. The presence of minerals under the ground or of distant lands was irrelevant as far as their survival was concerned.

Two resources that were visible and available in some quantity, depending on location, were timber and stone. Both were put to good use but timber had two big advantages over stone: it was easier to work, and it could be burnt. Since then, timber has been in constant use and it is almost impossible to exaggerate its importance in the development of human progress. It has continued to be used as a fuel, and in this role it turned grain into food, clay into pottery, ore into metal, seawater into salt and sand into glass.

Timber made transport possible. Until little more than a hundred years ago every ship was built of wood and at the end of each voyage they came alongside jetties or wharves which were also made of wood. Carts were made primarily of wood and they crossed bridges that were made of wood.

Wood was also the essential building material. Houses, large and small, could not be built without wood and often were built entirely of wood. Livestock were held in enclosures made of wood and paddocks were fenced with wood. Windmills were built of wood, and so were water wheels. Both were major sources of power before the use of electricity. Another, steam, was produced largely by wood, and entirely so in those places that lacked other fuels such as coal.

Tools and weapons had handles made of wood. Containers such as boxes and barrels were made of wood. Furniture was made of wood.

At any time during human history the progress of the human race would have come to a halt if wood had not been available. There was no other resource so readily available that could be put to so many important uses.

The availability of this essential resource, or the lack of it, determined the history of entire civilisations, which rose and fell largely because of the availability of wood. Wars were fought for it and land colonised for it. For much of human history wood, the essential resource, was as important as oil is today.

In his book *A Forest Journey*, from which this chapter is drawn, John Perlin retells the Epic of Gilgamesh, which describes how the forest was first conquered.

This took place about 4700 years ago in the city-kingdom of Uruk in southern Mesopotamia. The ruler of Uruk was Gilgamesh and he was determined to write his name into history. In order to do this he decided to extend the city and make it splendid, and for this he needed timber. Fortunately for him there was nearby a primeval forest that was so large that nobody knew exactly how extensive it was. It was almost impossible to penetrate and nobody had tried. In any case the chief Sumerian deity, called Enlil, had appointed a fierce demigod called Humbaba to protect the forest and the gods who lived there.

Gilgamesh's advisers warned him of the dangers, but he swore he would enter the forest and fell the cedars. So with a band of supporters he rode into the forest to rid it of Humbaba and thus make it available for cutting. They started to cut the trees and the noise woke Humbaba, who warned them to leave the forest alone. There was a fight and Humbaba, in spite of his ferocity, was killed.

Gilgamesh (that is, civilisation) was now the master of the forest and 'for two miles you could hear the sad song of the cedars'.

This print, made about 1600, shows trees being converted to usable timber.

Gilgamesh started to plunder the forest and when Enlil heard of it he put a curse on him: 'May the food you eat be eaten by fire; may the water you drink be drunk by fire'. Meaning that the consequences of civilisation plundering the forests would be drought and fire.

As a result of this and similar massive exercises in deforestation the irrigation canals silted up and the salinity of the soil increased. The production of crops declined and by 2000 BC the last Sumerian empire had passed. Three hundred years later the centre of power was Babylonia, which was not affected by salination, and the once mighty town of Sumeria had become a collection of unproductive villages.

So started, according to legend, a sequence that was to be repeated many times throughout history, and which is still being repeated. Civilisations rose on the abundance of timber. This was exploited without thought for the future and when the timber was exhausted the civilisation went into decline.

In Cyprus, for example, charcoal was used as a fuel to smelt and refine copper, which during the fourteenth and thirteenth centuries before Christ was in demand throughout the known world. The production of one ingot of copper needed six tonnes of charcoal, which came from 120 pines trees, which deforested nearly two hectares. For one ingot.

A Bronze Age shipwreck was found to be carrying 200 ingots of copper from that island. That cargo alone had cost nearly 24 000 pine trees. The production of copper deforested about ten square kilometres a year, and about the same amount was needed to supply fuel for domestic use and for minor industries such as pottery. All this on an island of only 9324 square kilometres. The result was inevitable.

Later, in Classical Rome, the consumption of wood was equally reckless. The sumptuous villas of the wealthy were centrally heated by wood-burning furnaces. In Germany a Roman villa of this type, now used as a church, was used to estimate the consumption of wood required to heat the building to an adequate level. It was found that the furnace needed 130 kilograms of wood *per hour*.

The climate of Rome might have resulted in a lower consumption there, but by the first century after Christ Rome was using wood for other purposes and on a much larger scale. The buildings which made Rome one of the wonders of the world demanded huge quantities of concrete made from lime, and lime was produced by burning limestone. One tonne of lime needed the fuel of an oak log half a metre in diameter and nearly ten metres long, or two fir trees about the same size. The requirement of one lime kiln was about 6000 kilograms of wood *per day*.

Not only did civilisations need wood to become established, they also needed a constant supply to maintain themselves. Eventually Rome had a fleet of ships doing nothing else but bringing wood from North Africa, Spain and France. The distant countries which supplied wood when local stocks had been used in turn became wealthy and powerful.

When Classical Greece had exhausted its own supply of wood it had to rely on the forests of Macedonia for further supplies. Macedonia, until then almost unknown, suddenly became rich and powerful. This power in turn became expansionist and the King of Macedonia, Alexander the Great, went on to conquer almost the entire known world.

Similarly, until about 2000 BC the island of Crete was as

insignificant as most small islands in the Mediterranean. But about that time its forests began to supply Mesopotamia, which had exhausted its own supplies. This trade turned Crete into one of the most powerful states in the known world.

Without huge supplies of wood the great civilisations of history would never have developed. The civilisations of Egypt, China, Greece and Rome were based on the availability of wood, and they went into decline when they were unable to secure wood from other sources or when the sources were so distant as to be impractical.

Wood therefore had a profound effect on the rise and fall of civilisations and this effect can be seen in very broad terms. But wood also had a very significant effect on the development of individual countries. While perhaps not threatened with extinction because of the lack of wood, its availability or the lack of it and the use to which it was put determined many political and economic decisions throughout their history. Many of these decisions, and the effect they had on the population, have receded into the past but they were vitally important at the time. In England, for example, wood determined much of its history from early times right up to the nineteenth century.

As an island trading nation, England needed ships for transport and defence. In Tudor times most of these were bought from foreign builders and in times of crisis extra ships were even hired from European ports. This was expensive and irksome because it made England dependent on others for trade and security.

Elizabeth I introduced a policy designed to increase the number of locally built ships and at no cost to the Crown. This policy restricted local sea trade to ships built in England and a major import, French wine, could be brought only in English ships. Similarly, a major export, fish, was made much easier if the ships were English. The result was that shipbuilding in England was made much more profitable, and the ships that were built could be requisitioned by the navy in times of need. In the five years from 1571 the commercial fleet increased by 7510 tonnes and by 1592 there were forty-two more ships of 100 tonnes or greater than there had been fifteen years earlier.

Successful though this policy was, it depended entirely on a huge supply of timber. And that meant mature oak, which shiprights believed was unequalled for the purpose by any other timber in the world.

The size of warships increased dramatically in the sixteenth century because of their need to carry heavy armament, and the consumption of oak, mostly from Sussex, increased accordingly. The

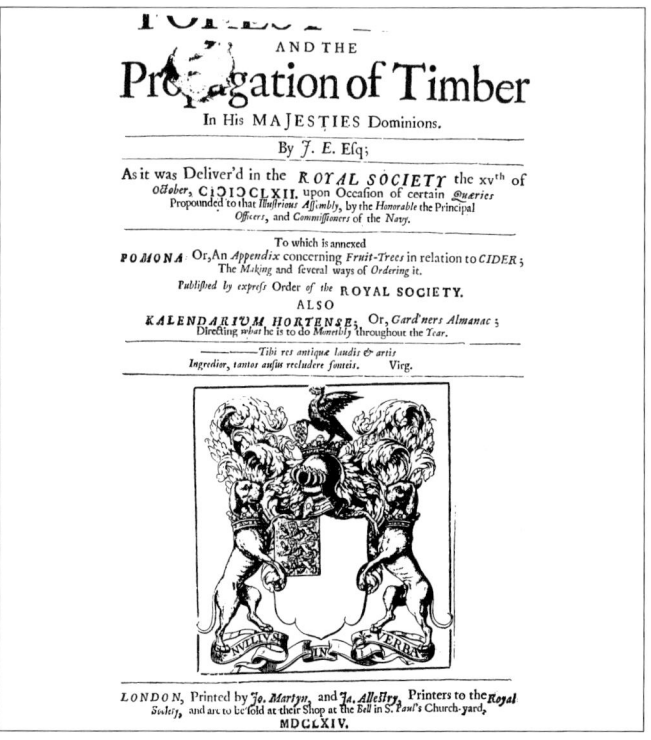

The title-page of John Evelyn's Sylva, *published in London in 1664.*

repair of four navy ships needed about 2000 tonnes of oak and this came from over 1700 oak trees. Building a large naval ship required about 2000 oak trees and each had to be over a hundred years old because younger trees did not have the required strength.

Others, too, were competing for supplies of timber. Leadminers in Derbyshire found over 90 000 large and small oaks in a nearby forest. Twenty-seven years later they had used about 93 per cent of them. Glassmakers also consumed wood at an alarming rate. Some used portable factories so they could move from one stand to the next, others bought woodland to guarantee supplies for a few years. Copper makers and iron smelters had an equally voracious appetite. In Surrey, two ironmasters alone used over 2000 trees in one year.

It was not long before the navy was complaining that it could not find sufficient good timber to maintain and develop its fleet. An Act was therefore passed to preserve mature trees for the navy within about 20 kilometres of the sea and within the same distance of major rivers. There were exceptions to this, however. Parts of Kent, Surrey and Sussex were excluded ostensibly because these areas were thought to be still well stocked. However, these areas also supplied fuel for ironworks, which were owned by some of the most influential men of the day. Sir Thomas Gresham, for instance, owned ironworks in Sussex as well as being chief financier to the government. Another, Viscount Montague, had been selected to sit in judgment of Mary Queen of Scots after she was accused of conspiring to kill Queen Elizabeth. Men like these, and there were many, were more than capable of protecting their interests against any Act of Parliament.

Local outrage eventually led to the passing of another Act which specifically prohibited the erection of new ironworks in these areas and also the felling of timber for making iron. Even so, this was done not so much to conserve trees but because by then the ironmakers had earned a nasty reputation by selling cast-iron cannons to potential enemies overseas.

At the same time the shortage of timber led to the use of coal as an alternative fuel. It was not popular and had even been banned in earlier times because of its objectionable smoke. Now this had to be tolerated because wood was too precious for the purpose.

The reckless use of wood, and its destruction when land was cleared for agriculture, caused much concern among those who could see what was happening. They also recognised that in spite of this heavy use none of the users were planting timber for future supply. Their gloomy predictions were largely ignored but by the early years of the reign of James I they had become obvious to everybody. In south Wales builders had to use stone and lime because there were no supplies of timber, and the Great Frost in January 1608 made those who had no access to coal realise how little wood remained for fuel.

The truth of the situation was brought home by an agricultural writer called Arthur Standish, who spent four years travelling around England assessing the remaining forests. He was appalled at the destruction he saw, and commented that the destruction was greater the nearer he approached London. Urging the need to replant, he commented that without wood there would be no kingdom.

Not surprisingly, this attracted the attention of King James, who became Standish's patron. James had himself become aware of the seriousness of the situation and had on more than one occasion called on Parliament to enact legislation to preserve forests, but all had been rejected. He had then used his royal prerogative and issued proclamations forbidding the use of timber as firewood and requiring new houses built within a mile of the suburbs of London to be built with brick or stone. Houses built of timber would be 'pulled down to the ground and . . . utterly destroyed'.

Unfortunately James' zeal for timber conservation was excelled by his ability to run up debts, which he did on an astronomic scale.

A foreign diplomat advised his government that James' debt stood at about five million pounds at a time when a working man earned about ten pounds a year and could, with prudence, save about a third of that. The diplomat also reported that in order to relieve this debt, James planned to sell some of the royal forests.

At first he planned to sell some of his more distant forests but public reaction forced him to abandon this plan. So instead of selling forests he decided to sell the wood instead. In 1612 he came to an arrangement with the Earl of Pembroke whereby he allowed Pembroke to build four blast furnaces and three forges in the Forest of Dean, which was the largest stand of oak in England. In return, Pembroke would buy a guaranteed amount of wood from the king for twenty-one years, which would provide the king with an income during that time of about fifty thousand pounds.

Once again there was a local outcry and some of the wood allocated to Pembroke was burnt before he could use it. The Privy Council intervened and Pembroke's privilege was withdrawn. This saved the Forest of Dean but did little for James' debts, which were inherited by his son Charles I.

With finances stretched to the limit Charles sent the navy to the West Indies to capture loaded Spanish galleons on their voyage home. But as this was an uncertain venture a commission was formed to examine other ways of increasing revenue. One of the methods they came up with was, again, to sell the royal forests.

Knowing the probable consequences of this they decided that the forests would be sold a little at a time to avoid the devastation that would have followed his father's venture with Pembroke. They decided to start with forests in the Midlands and Sir Miles Fleetwood, a courtier, was put in charge. He started by selling the Leicester and Feckenham forests and he was so successful that he then turned to the Roche and Selwood forests. Before long it seemed that any part of the royal forests was for sale provided the price was right. The Earl of Northampton, for example, not only bought rights to part of the Forest of Whittlewood but was also allowed to clear it and turn it into farmland.

Successful though this was at raising income, Charles soon faced a dilemma. He had a duty to maintain the navy but with no income forthcoming from a hostile parliament he had to do so from his private purse. If he bought wood from private owners he had to pay for it like anybody else and, worse, was faced with competition from other users. This competition was fierce and had already pushed the price of timber to record levels. The alternative was to draw timber

WHO NEEDS TREES?

Throughout most of human history, trade, exploration and warfare depended on ships made of wood.

from the royal forests, which he could do for nothing other than the cost of labour and transport.

The first to go were his forests in Hampshire, which were close to the naval depot at Portsmouth. By now the navy's need was so great that they marked 10 000 trees in one year, but two years later they were still unable to keep the fleet in repair. In order to keep them supplied Charles bought a thousand trees from the Earl of Southampton, who was selling them because his financial needs were nearly as great as the King's.

But Charles could not maintain the navy in this way and when the navy announced that it intended to build its first triple-decker warship he sent the surveyor general out to find timber in the royal forests. He came back with a list of forests which would provide the 2500 tonnes of timber the project needed. Although many of these forests had never been used they were now plundered for the navy.

While Charles and his predecessors managed, somehow, to keep the navy supplied with timber to build and repair ships, there was one naval requirement that could not be met from English forests. English trees were not suitable for that most important part of a ship: its masts.

GROWING UP

This print of 1717 shows workmen felling and measuring timber while the owner negotiates a price with the merchant.

Naval masts were imported from Scandinavia, Poland and Russia and on the way they had to pass through narrow shipping lanes that could easily be blocked by an enemy. Without masts, the English navy could not sustain a prolonged war and defeat would be inevitable.

This came to a head in 1658 when the Dutch threatened to block the channels from the Baltic and thus cut off the supply of masts to the navy. This was intolerable, if only because it confirmed that, in spite of its naval might, Britain was at the mercy of anybody with enough ships to mount a blockade. In the end a fleet of sixty British ships sailed into the Baltic and succeeded in keeping it open for this essential trade.

Eventually even the Scandinavian forests could not supply the timber needed by the navy's ships. As these ships became bigger so did the masts, and they were not to be found anywhere in Europe. Ships of the line, the majestic battleships of the day, were now 30 to 50 metres long, ten to fourteen metres in the beam and had a draft of five to six metres. This was not simply vain grandiosity. These ships carried anything from sixty to a hundred guns and had to provide a stable platform when they were in use.

Because natural timber could no longer be found to produce the masts needed for these ships, they were made by splicing smaller pieces together to arrive at the required size. The trouble was that they were not as strong as a single piece and in a warship that could be crucial. Masts were a prime target in any naval encounter because if they could be brought down the ship was robbed of its momentum. It might remain as a firing platform, but it was not going anywhere. Unfortunately, spliced masts were much more vulnerable than those in a single piece and could not be relied upon to remain standing in a storm, let alone during an intense engagement.

The problem remained without a solution until reports were received from the American colonies that some of the biggest trees in the world had been found in New England. The first English party to arrive there in 1602 recommended it as a place of settlement because it could supply masts to England without interference from enemies.

It was not until 1634 that the first shipload of New England masts was sent to England, but they lived up to all expectations. The Commonwealth Government of England, which had taken over after the murder of King Charles, decided to encourage the supply of masts from the colony and wrote accordingly to their 'loving friends, the Governors and Commissioners of . . . New England'.

This was the start of the transatlantic trade in masts which brought great relief to England, not only because of the freedom from interference but also because the size of the masts was almost exactly that required for the bigger ships. Ten or more ships arrived each year from New England and each ship had a cargo of between twenty and forty masts. What is more, they came from a British colony.

Inevitably, though, they were not free from interference. In the first year of the trade the Dutch took two ships carrying masts from

GROWING UP

This picture shows how heavily settlers in Australia relied on wood. It was the essential resource.

New England and eventually the navy had to provide an escort for the convoys.

Soon there was another kind of interference. The attempts by the British to reserve the best timber in the colonies for its own use met increasing resistance from the American settlers. In the end the settlers thwarted Britain's attempts to secure the best trees by a combination of cunning and violence. In his book, Perlin writes:

> The victory of local lumber interests over the wishes of the Crown to preserve the woods for masts hoisted the flag of independence so many notches higher. More than ever before, the people had supported the belief, noted by an English gentleman in his impressions of America, 'that His Majesty has no business in this country . . . [that] the country is ours not his . . . the property being ours so also are the masting [trees], and if we can sell them to others for more money we are at liberty, notwithstanding the two acts of parliament [protecting] those masting trees, as we have nothing to do with their country, so they have nothing to do with ours.'

This was a serious matter because by the beginning of the eighteenth century every ship of the line in the Royal Navy carried masts that had come from New England. And so when American Independence was declared in 1776 England was denied its supply of masts and was once again in great need.

WHO NEEDS TREES?

England was also denied a place to which it could send its convicts and it was largely because of this that the government decided to settle Australia as a convict colony. Some historians have suggested that the need for masts was no less pressing and that as Cook's earlier reports suggested that good masts could be found on Norfolk Island, together with flax for cordage, this provided all the justification needed for a settlement on the other side of the world. It is unlikely that this was the primary consideration, but it almost certainly added weight to the main reason, which was to remove the convicts from the prison hulks of Britain. Certainly Norfolk Island was settled by a party from Sydney within a few weeks of the arrival of the First Fleet when there was no obvious need to do so. In the end it was to prove disappointing for the navy. Most of the trees were unsuitable for masts, and there was nobody in the First Fleet who knew anything about flax.

Meanwhile, in America the settlers started to use timber at a furious rate. Many settlers relied on the sale of the timber they cleared to provide an early profit and this was sometimes enough to repay the cost of the farm. Indeed, they were encouraged to do so by those selling the land: 'Buy this farm, cut off the wood, haul it to market, get your money for it and pay for the farm . . . The owner estimated there will be five hundred cords of market wood.'

Nor was there any shortage of customers for wood. Railways and steamboats required great amounts. If the wood was too remote to be sold it was felled and burned. The ashes were mixed with water to produce a strong lye, and when this evaporated the lye turned into potash which was in demand for making soap and other purposes. The sale of potash often paid for tools and other necessities if nothing more.

Between 1811 and 1867 nearly five billion cords of wood had been used in industrial furnaces, in fireplaces, and by the railways and steamboats, and half of this had been used in the last seventeen years. Obtaining five billion cords meant the destruction of over half a million square kilometres of forest, an area nearly equal to that of Michigan, Illinois, Wisconsin and Ohio combined.

During the same period another 64 000 square kilometres of forest were cut to provide wood for building houses, ships, railways and many other equally demanding purposes.

What made both activities even more destructive was that the cutters generally removed the younger trees because they were easier to fall and remove. Thus, not only was a great deal of timber removed, but the future of the forests was jeopardised as well.

Yet another significant demand saw more than 80 000 square

GROWING UP

Timber framing for a modern Australian house. Most of the timber for framing now comes from plantations.

kilometres of timbered land cleared for cultivation between 1850 and 1860. Livestock allowed to graze in forests also did considerable damage by eating seedlings and by removing bark. The need to improve forest grazing led to burning, which was done for exactly the same reasons as it was by Australian Aborigines. It also had exactly the same result: forests became open woodlands with a luxurious growth of grass.

In 1850, 25 per cent of the land of the United States was covered with forests. By 1870 this had been reduced to 15 per cent.

Obviously much of this had been necessary. Land had to be cleared for farming, and wood was needed to fuel transport and industry. But it had been done with no thought for the future. In 1882 N. Egleston, a leading authority on forestry, wrote in an issue of *Harper's Monthly*:

> We are following . . . the course of nations which have gone before us. The nations of Europe and Asia have been as reckless in their destruction of the forests as we have been, and by that recklessness have brought themselves unmeasurable evils, and upon the land itself barrenness and desolation. The face of the earth in many

WHO NEEDS TREES?

instances had been changed as the result of the destruction of the forests, from a condition of fertility and abundance to that of a desert.

After countless generations the curse on Gilgamesh was as effective as ever.

By now the need for wood was so obvious, and its destruction so serious, that attitudes had to change. There were centuries of history to warn of the consequences of destroying the forests. Entire civilisations had fallen as the forests had receded. And they had receded because the supply of wood was assumed to be limitless. The forests were so vast that it was thought they could supply wood for ever. So there was no urge to preserve them or to replant them for future use.

In spite of the compelling evidence to the contrary, this view was held for centuries and right up to modern times. By the time Australia was being settled by Europeans there was enough history to warn of the dangers of destroying forests.

It was ignored.

3 The Limitless Resource

When the First Fleet arrived in Australia in 1788 its company of convicts and guards found a land that offered practically nothing for their survival. In spite of the earlier reports by Captain Cook and Joseph Banks, who had described lush meadows and running water, they found a barren place with harsh, stunted growth.

There was no agriculture, no milk-producing animals and no food that they recognised apart from fish, which were difficult to catch. All the food that is grown in Australia today apart from the macadamia nut has been introduced since European settlement.

Certainly the Aborigines lived off the land. As food gatherers and hunters they found that the land usually produced enough to meet their needs. But the food that sustained them was not palatable to the Europeans, nor could it support the thousand or so people who arrived with the First Fleet in a small area of settlement. The Aborigines survived because they could move from place to place in search of food. But a nomadic lifestyle was hardly compatible with a convict colony, which was the purpose of this settlement.

The settlement faced a bleak future from the first day. Having landed at Botany Bay, and finding it a poor anchorage with no food or fresh water, they realised that they were entirely dependent on the supplies they had brought with them. Had Phillip not found the more promising site at Sydney Cove a few days later there was every likelihood that the settlement would have failed.

THE LIMITLESS RESOURCE

The base of a red cedar tree, photographed in north Queensland in 1967.

On 26 January 1788 Phillip's advance party went ashore at Sydney Cove while the ships of the fleet made the short voyage from Botany Bay. The Tank Stream, which was the only fresh water they had found, ran into Sydney Cove through trees that grew to the edge of the Cove. This was to be the site of the first settlement and on that day the advance party started to clear the ground.

Soon the sound of axes striking trees and the noise of them crashing to the ground rang around the cove and ended a tranquillity that had lasted since time immemorial. It was not to return.

Early attempts at growing food and raising livestock from those they had brought with them were not successful until fertile land was discovered at Parramatta and along the Hawkesbury River. Until then, and for some time after, the Europeans were close to starvation. Salvation could come only with the arrival of the next fleet but they had no knowledge of when it would arrive, or even if it had left.

Because the entrance to the harbour could not be seen from Sydney Cove a flag pole was erected on South Head which could be seen from the settlement. It was constantly manned and a flag was to be hoisted when ships were sighted. Meanwhile the settlement had become lethargic, rations had been reduced to the bare minimum, and there was no future. Until 3 June, when the flag was raised. Captain Tench wrote:

I was sitting in my hut, musing on our fate, when a confused clamour in the street drew my attention. I opened my door, and saw several women with children in their arms running to and fro with distracted looks, congratulating each other, and kissing their infants with the most passionate and extravagant marks of fondness. I needed no more; but instantly started out, and ran to a hill, where, by the assistance of a pocket glass, my hopes were realised. My next door neighbour, a brother-officer, was with me; but we could not speak; we wrung each other by the hand, with eyes and hearts overflowing.

The arrival of the Second Fleet brought almost as many problems as it solved, but it also saved the faltering new settlement.

Later, when land grants were given to officers and time-expired convicts the tree clearing that had marked that first day at Sydney Cove became more extensive. There was still an urgent need to establish crops and raise livestock and both required open ground.

Some historians have since criticised this land clearing by these and other settlers because they saw trees as 'useful' and little else. One historian complains that a particular settler described 'the trees in terms of their usefulness for cart building, roofing, slab hut and stockyard construction' and goes on to say that 'various historians have commented on these narrow, purely utilitarian attitudes to the forest environment'.

Although there may later be some truth in this, it overlooks the fact that the earliest settlers were concentrating entirely on their own survival. If they were to survive they had to bring land into production for crops and for grazing. They had to build houses for themselves, they had to build yards and they had to clear the land. So it is not surprising that they saw trees as useful and little else. Trees were the only resource available. Other considerations could not surface until survival was assured and a more leisured class had emerged, which was still some time ahead. Nor was the use of trees a result of a sudden awareness of their value. This was a British settlement and the new settlers were from a country that had used trees for centuries.

Although Australia was not as heavily forested as is often supposed, the early settlements, which were on the south-east coast and in Tasmania, were areas that were well forested and offered a seemingly limitless supply of timber. So much that it seemed impossible that it could ever be exhausted.

Many forested areas consisted of eucalypts but it was not long before a more valuable and important timber was found. This was red cedar and its discovery marked the start of the timber industry in Australia. Red cedar was important because of its rich quality and

THE LIMITLESS RESOURCE

Springboards and axes being used to fell trees in a Queensland rainforest. Place and date unknown.

it was felled for its inherent usefulness as timber rather than to clear ground for other use.

Cedar was probably first seen in 1790 by Governor Phillip during a journey along the Hawkesbury. He described it as being similar to a large walnut tree, deciduous and bearing a small fruit. The cedar is similar to the walnut and it is deciduous, but it does not bear fruit. But as cedars were later found in large numbers in that area it is probable that Phillip's tree was a cedar.

In 1795 a cargo of cedar from the Hawkesbury was sent to England in the hope that it might prove to be of value and this was almost certainly the first export cargo to leave Australia. The early settlers on the Hawkesbury soon developed a trade in cedar and this resulted in an early attempt to control it. On 2 April 1802 a General Order was issued:

> It having been represented to the Governor that some of the settlers at the Hawkesbury are making a traffic of the cedar growing on, or about that river, he strictly forbids any cedar being cut down

but by his permission or to the officer commanding at that place, and if any cedar logs or planks are brought from any part of that river, or any other settlement without the Governor's permission, such logs or planks will be seized for the purposes of Government, and the boats or carts containing them confiscated for public use.

Cedar was also found on the Hunter River soon after its discovery in 1801 and before any settlements were established along it. When the settlement of Newcastle was formed many convicts were employed in cedar getting. Between March 1805 and August 1806 over 150 000 feet of cedar were shipped, of which two-thirds were privately owned and the rest owned by the government. The trade in cedar was now well established and it was frequently advertised as being available from Sydney dealers.

The next discovery was at the Illawarra on the coast south of Sydney. The first shipment arrived in Sydney from the Shoalhaven River in January 1812. By 1819 this trade was so extensive that again the government moved to control it. An order issued that year stated that anyone found cutting, removing or possessing cedar from Appin and Illawarra would be prosecuted.

The following year it became possible to cut a specified amount of cedar provided permission had been granted and considerable quantities were taken. As a result, in June 1826 a further order was issued recalling and cancelling the permits that had previously been granted. There was a strong reaction to this but it was defended in the *Sydney Gazette* on 1 July 1826. The reason, it said, was that the export business would soon be ruined if cedar-getters were to continue destroying timber and, equally important, 'it was stated that vice of the most abominable kinds was practised amongst those cedar hordes, to the total annihilation of every correct principle'. Having had their existing permits cancelled the cedar-cutters applied for new permits and the numbers were so great that the Surveyor-General was sent to the Illawarra to examine the situation at first hand.

He found the cutters working in thick scrub but in relatively small areas that were rich in cedar. Which stands they worked was largely determined by the ease with which a road could be built in order to remove the timber. The scrub was so thick that only three or four trees could be cut at the same pit. This had produced a code of ethics among the cutters. A pair of sawyers could claim only those trees that could be sawn at one pit. If they felled trees that required a new saw pit they had no claim to them and anyone could cut them once they had built a pit for the purpose. This code was simple and well understood and there is no record of any dispute arising from it.

There was also a structure among those involved in this trade. There were the traders in Sydney who employed sawyers to produce the timber. The traders then organised transport and the cedar was often brought to Sydney in their own ships. Then there were the cutters themselves. They lived in the bush and worked either for the Sydney traders or on their own account. Finally there were the carters who brought the timber from the pit to the loading places. The main one was at Kiama, where there might be six ships at any one time loading cedar or unloading supplies for the cutters.

Cedar cutting in the Illawarra flourished through the 1830s and then started to slow down. By 1850 most of the accessible timber had been cut out and the trade came to an end, although it lingered on in Kangaroo Valley.

Meanwhile, cedar was discovered along the coast north of Newcastle. After small cuttings had been made at Port Stephens, in 1821 Solomon Wiseman was given permission to cut 10 000 feet there. By 1823 the cedar trade at Port Stephens was well established and ships loaded there with cargoes for England.

Settlers discovered good stands of cedar as they ventured further north. On 26 June 1823 the *Sydney Gazette* reported that the *Surrey* at Port Macquarie had nearly finished loading one of the finest cargoes of cedar ever seen. Further discoveries were made along the Hastings River. By the end of 1824 one party was known to be working about 40 kilometres from Port Macquarie and cedar was found in abundance in the tributaries of the river.

By 1828 the trade had extended to the Manning River. One settler had cut about 200 000 feet since taking up land on the river, and of this about 60 000 was still waiting shipment. The trade was supported by fresh discoveries along its tributaries and in 1842 the forests along the river banks were still abundant with cedar, rosewood and almost every type of hardwood.

In February 1836 the *Gazette* reported 'stupendous cedar' on the Macleay River and that a number of ships were waiting cargoes. By March 1841 over two hundred sawyers were working on the river but by September the following year the accessible timber had been cut out and only a dozen sawyers were left. On 12 October 1842 the *Sydney Herald* reported: 'There has been nothing but one wholesale system of plunder in the cedar trade carried on on the Macleay for some time past and some of the purchasers of stolen cedar were men pretending to respectability.'

By 1842 cedar had been discovered on the Nambucca and the Bellinger. There, the first tree cut was about 90 feet long and produced about 20 000 feet of cedar. By 1845 there were twenty pairs

GROWING UP

A bullock team hauling timber from the forest at Brooloo in 1924.

of sawyers on the Bellinger and their output was estimated at two million feet per year.

Cedar was also discovered on the Richmond River in 1842 and after a small start the river produced cedar for many years. In 1859 there were thirteen ships loaded with cedar waiting at the river mouth for a favourable wind. In 1857 the *Sydney Morning Herald* reported on the Richmond, where it had asked the sawyers how long they thought the trade would last.

> Some say ten or fifteen years—some say the present generation will not see the end of it. It is clear that even now the sawyers spread out a long distance to cut it, and many instead of fixing their wives and children on the spot have to leave them at the head of the creeks whence the cedar is floated down and go to their work three or four miles or perhaps further. They are also continually forming new camps where they find a fresh lot of cedar trees . . . It must be a fine trade for those who are in it, for the sawyer generally spends a third of his time over the keg, and there is not the slightest difficulty in getting it.

From there the trade moved to the Tweed, where it was well-established by the 1860s.

THE LIMITLESS RESOURCE

Queensland had become a separate colony in 1859 and one of the first acts by the new government was the introduction of the first timber regulations. Licences were needed to cut timber on vacant crown land and on pastoral leases, but not on freehold land. There were two types of licence: one to cut any kind of timber including cedar, hardwood and pine which cost four pounds a year; and one to cut only hardwood for two pounds a year. By 1861 there were 125 licenced timber cutters and some were working as far north as Gladstone.

Later regulations required pine logs to be removed within three months of felling and other species within twelve months, after which time the logs would be confiscated. This was an attempt to avoid the waste that had been so common among cedar-cutters further south, but it had little effect. Lack of buyers, lack of floods to take the logs downstream, or country that was inaccessible meant that trees that had taken centuries to grow were cut down in a matter of minutes and then left to rot. Henry Massie, a crown bailiff, told the Committee on Forest Conservation in 1875 of the problem: 'By parties cutting timber and allowing it to lie; there is no restriction; they take a licence and go out and cut, and what suits them they take away.'

By the 1870s the rainforest on the Atherton Tableland along with the forests of the Daintree and Mossman valleys were found to be well supplied with cedar. By 1877 most of the cedar had been cut from the Daintree and Mossman. In 1878 the *Port Douglas Times* reported that in the previous four years fifty-nine ships had taken away about 4¾ million super feet of cedar, and that was a conservative estimate of the total that had been taken from these two valleys. During that year a million feet of cedar was waiting shipment from the Mulgrave.

By 1881 cedar was being cut in huge quantities along the Herberton, Butchers Creek, the Mulgrave River and from Ravenshoe. A group of cutters and merchants had examined the Barron River and had decided that cedar could be sent down the river even though it was nearly one hundred kilometres long and was interrupted by the vertical drop over the Barron Falls. They thought that the basin at the bottom of the falls was deep enough to avoid damaging the logs and that losses would be slight. Although there were some spectacular mishaps they were proved generally correct and the use of the falls was fairly successful in removing timber from the Atherton Tablelands. It must have been an amazing sight.

In spite of the fact that cedar was cut from the Illawarra to the Daintree the cutters themselves seemed to have much in common. They worked in pairs, they worked deep in the bush and their habits appalled contemporary society. John Henderson, in his *Excursions and Adventures in New South Wales*, wrote:

> These sawyers and their mates are a strange wild set, comprising in general a good proportion of desperate ruffians, and sometimes a few runaways, they themselves commonly being ticket of leave men or emancipists . . . their fare is but simple, consisting only of salt beef, damper, tea and sugar . . . They are certainly the most improvident set of men in the world, often eclipsing in recklessness, misery and peculiarity of character the woodcutters of Campeachy and the lumberers of the Ohio and Mississippi . . .
> When rum is brought to these abodes of labour and wretchedness, and a few sawyers are convened, then begin the scenes of riot and mischief.

There are many similar descriptions of cedar-cutters and there is no reason to doubt them. However, they are also true of other people who worked in the bush in the nineteenth century. Stockmen, shepherds, shearers and many others were described in similar terms. This is not surprising because they all had much in common. They all worked in remote areas, they lived in communities that were nearly always without women, they worked hard and earned well, and when they had a chance they would attack grog with a ferocity that appalled most urban observers.

There is no doubt that cedar-cutters were hard working, whatever their other failings might be. Most were working for themselves rather than for wages and there was nothing to be gained by idling.

The work consisted of two separate activities: felling trees and then cutting them into boards. Felling was the easier of the two, as little concern was given to where the tree might fall or what damage it would do so long as it missed the pair doing the cutting. After the tree was felled the top foliage was removed to leave a clean log, and the log was cut into lengths that were easier to move and would fit the length of the saw pit. However, a great deal of cedar was exported as logs and therefore not cut on site.

The saw pit consisted of a deep trench with a timber frame over it to support the log. Once in place, the log was marked with a straight cutting line. To do this, the sawyer and his mate stretched a string covered with chalk along the log. When the string was in the correct place they snapped it onto the log so that it left a clear and straight white line.

The sawyer stood on the top of the log and guided the saw along the line. His mate was on the other end of the saw, below the log, in the bottom of the pit. The saw was about 1.8 metres long and

THE LIMITLESS RESOURCE

Loading logs onto a wagon.

wider at one end than the other. The sawyer on top used the wider end and pushed the saw downwards in the cutting stroke. His end of the saw was fitted with an extension called a tiller so that from his position on top of the log he could use the full cutting edge of the saw.

His mate below, walking backwards and covered in sawdust, pushed the saw back up for the next cut. His end of the saw was fastened to a block of wood called a box. This could be tapped clear of the saw if it had to be removed from the log. Each cutting stroke advanced the saw between five to ten millimetres depending on the hardness of the wood. The teeth were never perfectly set and they left a regular pattern of marks on the board.

Although convention restricted a pair of cedar-getters to those trees that could be worked at the same pit, others were not so restricted and often had to move the logs a considerable distance to the nearest pit or to a loading point if they were to be shipped as logs. There were various ways of doing this. They could be sent down a swollen watercourse, as at the Barron River, or they could be sent down a hillside on a chute, also called a shoot and which still exists in many local place names.

This was a track which the log slid down. They were often simply scrapes on the ground but this was not highly regarded because the log collected stones and gravel on the way which made the later sawing more difficult. It was better if the chute were lined with timber. In either case the log was snigged to the top of the chute and moved with handspikes so that it went down end first.

Bullocks were also used to move logs, but driving bullocks was a specialised job and was usually done by a separate contractor. Moving the logs to a pit or loading point was called snigging. A chain was passed around the log and the bullocks simply dragged it along the ground end first so that it did not get jammed against standing trees.

Logs were also loaded onto bullock-drawn vehicles, although this was done only if the logs had to be moved over a long distance. There were two types of vehicles. The two-wheel jinker was difficult to load and was used only for short journeys with large logs. The other had four wheels and was called a top loader. When loading, the driver positioned the vehicle alongside the logs and about three metres from them. He then placed a pair of skids from the ground to the tray of the wagon. These skids were shaped so that they were a good fit over the top of the wheels. He moved the bullocks to the other side of the wagon and with the use of chains they pulled each log up the skids and manoeuvred it onto chocks that had been placed on the tray to receive it.

The country that attracted timber-cutters attracted hardly anybody else. The best timber was found where the forest was at its thickest and cedar, which was not evenly distributed within the forest, was no exception. Simply getting to it was a major operation. The ground was often very steep, the forest dense and there were no tracks.

Charles Archer wrote a letter to his father in Norway in 1844 in which he described the country around Durundur:

> Such country is of course utterly impracticable for a horseman, and not to be rashly ventured upon even on foot. I have myself entered, a very decently dressed bushman, and returned in almost a state of nudity—my shirt and other garments torn into strips which fluttered gracefully (like so many streamers) in the breeze. Happy is he who escapes with a sound back and arms, which on some occasions are lacerated as if the unfortunate might have done battle with an army of cats.

In fact, the country he described would have seemed tame to most cedar-cutters.

Not all timber-cutters looked for cedar. There was a huge demand for timber of almost any kind for building, fuel, for making carts and boats and for many other uses. Most timber for these uses could be obtained more easily than cedar, albeit without the profits that cedar offered.

Pit sawing in the bush. Note the tiller on the top of the saw which allowed the full length of the saw to be used with each downward stroke. At the lower, narrower end of the saw is the box, which could be knocked free if the saw had to be withdrawn before the cut was complete.

Much of this timber came from forests that were cleared for settlement, but the rest came from standing forests that were not being cleared. There was almost no control over this activity other than occasional government orders that were virtually impossible to enforce. One early order was rather unusual, however, and it was enforced with a fair degree of success. It was issued in Sydney on behalf of Governor Gipp on 14 April 1842, before Queensland became a separate colony.

> It having been represented to the Governor that a District exists in the Northward of Moreton Bay, in which a fruit-bearing Tree abounds, called Bunya, or Banya Bunya, and that the Aborigines from considerable distances resort at certain times of the year to this District for the purpose of eating the fruit of the said Tree: His Excellency is pleased to direct that no Licenses be granted for the occupation of any Lands within the said District in which the Bunya or Banya Bunya Tree is found. And notice is hereby given that the several Crown Commissioners in the New England and Moreton Bay Districts have been instructed to remove any person

who may be in the unauthorised occupation of Land whereon the said Bunya or Banya Bunya are to be found. His Excellency has also directed that no Licenses to cut Timber be granted within the said Districts.

This order was revoked in 1860 when the newly formed colony of Queensland issued the first timber regulations which have been described earlier.

These regulations, which required timber-cutters to be licensed, were little more than a means of raising revenue and did little to protect the timber. They were also almost impossible to police, so that the number of timber-cutters at work was little more than conjecture and bore little resemblance to the number of licences issued.

In 1864 the Queensland Government introduced a Special Timber Licence in addition to the two that were previously issued. This special licence gave the holder exclusive rights to cut and remove timber from a defined area at a fee of twelve pounds a square mile. The need for such a licence had been felt in the south-east of the state, where increased clearing for settlement was driving cutters into more inaccessible areas, which meant they had to face expensive costs in creating tracks which could then be used by others. Again, there was almost no means of enforcing that exclusivity.

Standing timber within two miles either side of a surveyed railway line could also be cut and in 1868 the *Crown Lands Alienation Act* was passed which allowed the transfer of large areas of native forest to private ownership. There was, in effect, practically no restriction on cutting timber or clearing land, and the restrictions that did exist could rarely be enforced.

The income from timber was also very important to this new colony. In 1860, the year after separation, timber exports (to New South Wales) were worth £2442 and timber was the eighth leading export. Imports that year exceeded exports by £2300 and this trade imbalance put a severe strain on the colony's finances. By 1869 exports had reached £30 000 and ten years later it had climbed to an impressive £74 007. Imports fell substantially during the same period. In 1876, for example, imports were down to about £4000. With an income as welcome as that, few politicians were likely to vote it out of existence.

The state's timber had become a very valuable resource. It had been cut mercilessly but what did that matter? Even a casual observer could see that apart from cedar there still seemed as many trees as there ever had been.

4 Forever is a Long Time

Although Queensland's forests were being cut and cleared on a large scale there was very little concern at any level of society. Most of the Great Dividing Range was heavily timbered and there was a general belief that it would last forever.

There was also no concern about clearing forests for settlement. The state needed people, and people needed farms. Land that supported forests was thought to be a waste of fertile soil. The fact that it was supporting forest was sufficient proof to most people that the soil was fertile, and it should therefore be put to 'better' use.

There was nothing new in this approach. The early settlement of America had proceeded in much the same way. The difference was that by the second half of the nineteenth century, when Queensland was being settled, there was enough evidence from America to show the consequences of deforestation.

The problem was that although trees seemed to exist in profusion the image was misleading. Trees are big and they occupy a lot of ground, but there are not as many as one might think. A one hectare paddock of wheat will today contain about 350 000 stalks of wheat. But one hectare of native forest will contain only about fifty trees, and not all of them will be useful for timber. A forest containing 350 000 trees, the equivalent of a hectare of wheat, will cover about 7000 hectares and seem enormous. Each tree will certainly be more productive than a single stalk of wheat but the sight of 7000 hectares

Native forests were cut for timber without any thought for the future. This advertisement, typical of many, appeared in the Dalby Herald *of 19 December 1868.*

Dalby Herald, December 19, 1868.

Cheap Timber

NOTICE TO THE PUBLIC

THE DALBY SAWMILLS having been removed to the Bunya Mountain, the Undersigned is now enabled to supply, in any quantity, a superior sample of TIMBER, and at such moderate prices as will merit public support, independently of any claims arising from the re-opening by local industry of the valuable sources of Building Material which these mountain forests contain, as the following figures will show: –

Pine Scantling, any size, reduced to superficial measurement, at 14s. per 100 sup. feet.
Pine Boards, 12 x 1, at 15s. per 100 sup. feet.
Pine Lining Boards, 9 x ⅝, at 13s. ditto ditto.
Pine Flooring Boards, 6 x 1, at 14s. ditto ditto
Pine Weather Boards, at 13s. ditto ditto.
Pine battens, 4 to the foot, at 14s. ditto ditto.
Cedar, at 25s. per 100 sup. feet.
Beech, at 22s. ditto ditto.
Deep Yellow Wood, at 25s. ditto ditto.

THE ABOVE PRICES ARE FOR CASH ONLY.

N.B. — The DALBY TIMBER YARD is at present in Scarlett street (Dr. Wuth's late residence, where all orders are to be received, and a good supply of well seasoned Timber will be always on hand.
Country orders will receive prompt attention.
ALEXANDER McDONALD

---

of trees will lead most people to think that the forest will supply wood forever without needing the slightest attention.

Until 1900 the control of Queensland's forests was in the hands of the Lands Department. In practice this was not control so much as enforcing the regulations that applied to timber-getters and issuing the necessary licences. This approach had nothing to do with conserving the forests. It was about collecting revenue, small though it was.

The regulations drew frequent criticism, but only on the grounds that they were often irksome to those who had to comply with them. For example, in 1864 William Pettigrew wrote to the Secretary for Lands and Works:

> By the present laws any person holding a timber licence can go to any vacant Crown Lands within certain limits, and take what timber he may choose, and any other licensed person can do likewise. As long as timber was plentiful, and is cut only by sawyers with the pit saw, this system wrought pretty well, but now that timber of all sorts has become more scarce—pine having to be drawn about six miles, and cedar sometime ten miles upwards—this system is unjust in its operations and will have to be modified.

His complaint was that licence holders going into new country bore the cost of establishing tracks and building bridges in order to bring the timber out. When timber was plentiful other timber-getters accepted that these facilities were for the exclusive use of those who had built them. But as timber became more difficult to find this convention broke down and others used the facilities to remove timber that the original cutters thought was rightfully theirs.

The Lands Department, then, was simply responsible for administering the existing legislation and recommending changes to legislation if it thought, as it did in this case, that there were legitimate reasons for doing so. The department was not concerned with forestry as such, and neither was anybody else.

Far from showing any concern about forests, other legislation actually threatened them. The *Crown Land Alienation Act of 1868* allowed people to acquire vacant crown land for settlement, and large areas of forest, especially in northern Queensland, passed into private ownership. The forest might be left standing or it might be cleared, but the decision was no longer made by the government.

What concern there was about the destruction of the state's forest came from a small but articulate minority who voiced their concern in newspapers and magazines. The *Queenslander*, a weekly magazine, published many articles describing what had happened overseas as a result of forest clearing and warning that Queensland might have to face the same consequences. But the discovery of vast areas of timbered country in North Queensland at about the same time seemed to guarantee timber supplies 'forever' and any anxiety about the future seemed unnecessary.

The declaration of the first Timber Reserve in October 1870 might indicate that the government was at last showing some foresight, but it was not. These areas were reserved so that the timber from them could be used exclusively by the government for public works, and especially railways. Although many Reserves were declared for this purpose most were revoked after the timber had been used.

It was also in October 1870 that concern for the forests was expressed to the government by a rather unlikely body, the

Acclimatisation Society of Queensland. These societies had become common in Australia in the 1860s and were supported by some of the leading citizens of the day. Their purpose was to introduce to Australia plants and animals from overseas so that they could develop naturally in a new environment. The landscape would be enriched and settlers would have pleasant reminders of the land they had left behind.

The societies had an almost missionary zeal and went to endless trouble to import new species. They were not very selective and were prepared to bring in almost anything so long as it was alive. In the process they introduced the Scottish thistle, Bathurst Burr and the blackberry. These plants, and other imports they made, went out of control in their new environment and have been serious pests ever since.

In October 1870 the Queensland society wrote to the Colonial Secretary voicing concern about the loss of forests and particularly the effect it might have on the local climate by reducing rainfall. The society suggested that statistics on land clearing be included in meteorological records. This was immediately rejected by the Registrar-General who thought he had enough work without that.

The society raised the matter again in 1873, when its annual report for the previous year included a section called 'Protection and Regulation of State Forests'. This was sent to the Premier and he passed it to the Secretary for Public Lands, whose response was more encouraging. He told the society that he would be willing to consider their suggestions on conserving forests.

On the basis of this, the society called a special meeting in May 1873 on the subject of forest conservation. Two papers were presented at this meeting. One, by the Gold Fields Commissioner at Rockhampton, John Jardine, dealt with the overall concept of forest conservation as it might apply to Queensland. The other was by the President of the Society and Clerk of the Legislative Assembly, L. A. Bernays, who described the practicalities of introducing forest conservancy to the state.

These papers are significant because they were the first to examine the consequences of deforestation in Queensland. The difficulty was that their concern seemed so preposterous to the public that members of the society were regarded as eccentrics to be pitied or scorned. Bernays wrote later: 'The bare idea of a scarcity of timber in Queensland, is often met with a smile of pity, or even a laugh of derision. One is asked to get upon any eminence, in any district, and to put to oneself the question—whether the sea of timber which meets the eye can ever run short.'

The government did nothing. Indeed it probably would not have

Logging at Eumundi in 1915.

given the forests another thought had not the Secretary of State for the Colonies the following year sent a questionnaire asking for information about the forests and the timber industry. The information was collected by Walter Hill, the head of the Botanic Gardens in Brisbane. He was in no doubt of the damage being caused and the poor state of Queensland's forests. He complained about destruction, waste, the lack of reserves and of any system to preserve the forests, let alone manage them.

Again the government did nothing. Instead it was left to a newly elected parliamentarian to take the initiative. In June 1875 John Douglas, the member for Maryborough, proposed that a select committee be formed 'to consider and report upon the best means to be adopted in order to preserve and promote the growth of Timber Trees, and to conserve Forests for useful purposes'.

The committee was duly formed and it reported later that year. It was not a very hard-hitting report and in many ways it was a disappointment. But it did draw attention to the fact that significant amounts of forest were being transferred to private ownership, and that the income, especially from northern cedar, was largely going to southern states. The *Queenslander* later reported:

> No Queenslander receives any part of the profits of the trade, and the timber-getters, in the north at least, have not been the pioneers of settlement. When the timber camps on the Daintree are abandoned, there will be no settlers left behind. We have in fact sold the magnificent cedar forests on the northern rivers to Victorian speculators for the paltry sum received as licence fees, whilst their men were at work falling the timber, and a very bad bargain for the colony it has been.

The report did suggest that a Forest Conservancy Board might be formed to help the minister but went on: 'It is probably too soon to expect that Parliament would authorise the appointment of a Conservator of Forests at the head of a department, though it is far from certain that such an outlay might be justified.'

The report was adopted by the Legislative Assembly and after that the government did nothing. Indeed, the Premier, A. Macalister, was strongly committed to closer settlement and thus the clearing of land. The same year that the committee sat Parliament passed the *Crown Lands Amendment Act* which relaxed some of the conditions of selection in order to encourage more people to take up land.

In spite of this activity in the late 1870s, by 1880 nothing of significance had changed. Regulations were made and amended, licence fees were collected, reserves were declared to meet specific government needs, and that was all.

In government terms, forestry lacked urgency. The general view was that there was still ample timber to meet the foreseeable demand. If there might be a problem in the future this was of little concern to most politicians. Nevertheless, once the subject of forest management and preservation had been raised it refused to go away. Newspapers kept returning to it and public attention, though never great, was maintained.

In 1880 the Under-Secretary of the Lands Department included in his report to Parliament a section on forestry and this became an annual feature. In that first year he wrote:

> The question of Forest Conservancy is occasionally brought under the notice of the public, but no result follows further than a general acknowledgement that the most valuable timbers of the colony are fast disappearing and that something must be done to stop the waste that is going on. I had anticipated that long ere this the

subject would have been dealt with by Parliament, but other matters of greater interest for the moment have occupied attention and forest conservancy is left for some more convenient season. At present the only thing which has been done in the direction of forest conservancy is to set apart some areas of land as timber reserves, which it is presumed will ultimately become State forests. This is, however, by no means certain, as efforts are continuously being made to have these areas declared open for selection by persons desirous of acquiring land, and sometimes these requests are backed by Parliamentary influence.

There was by now a clear need for legislation and there was enough concern in a generally apathetic public to support such legislation. But the need failed to rouse the interest of politicians, and the last sentence in the Under-Secretary's report indicates one reason why. People wanted to clear land for settlement with no thought for the future. Most politicians supported this as a desirable sign of initiative and capitalist enterprise, both undisputed virtues at that time. Politicians did not hesitate to encourage those who were prepared to put money into the land, and if this encouragement increased the politicians' stature among influential voters then so be it.

Politicians were also aware that the export of timber was a major contributor to state revenue. Passing legislation to conserve forests was likely to reduce this considerably. The lack of legislation, however, meant that the forests were undervalued. In short, there were too many conflicting interests. The demand for land was simply far too strong and the occasional articles in the press and the section in the Under-Secretary's annual report could be safely overlooked, or considered, commented on and forgotten.

Although concern for forestry was largely disregarded it did not entirely disappear. In 1889 R. M. Hyne raised the matter again in the Legislative Assembly and called for the creation of a Department of Forestry. Hyne was the member for Maryborough, where he had established one of the most important sawmills in Queensland. His concern for the future of the forests was therefore closely related to his concern for the future of his own business. At the same time he was also in a position to see the consequences of deforestation more clearly than most and he had considerable support in the House.

In the annual report of the Lands Department for 1889 District Surveyor McDowall made a similar point and in terms which were more aggressive than was usual:

> It can hardly be questioned that the time is approaching when the wholesale destruction of timber in many parts of the Colony—much of it of a wantonly wasted nature—will be severely felt.

GROWING UP

A logging camp at Atherton about 1910.

> Suddenly, when the depredations of a careless population have produced the inevitable result, the subject of forest conservancy will assume a prominence not yet accorded to it, and it will be a matter of general wonder that our shortsightedness did not allow us to realise that destruction without replenishment must lead to scarcity.

The government did nothing. The funds set aside by the government for forestry purposes in 1891 amounted to £65.

The need for a forestry department was spelled out every year by the Lands Department and in 1898 the report left no doubt that the time had arrived and it could be put off no longer. Faced with the inevitable, and after talking about forestry for thirty years, the government at last decided that a forestry department should be set up within the Lands Department. This department was formed in 1900 and it was the first attempt to bring order into what by now was chaos.

The first head of the department, with the title Inspector of Forests, was L. G. Board and he took up this new position on 1 August, having previously been Land Commissioner for the districts of Gympie, Maryborough, Bundaberg and Gladstone. He was assisted by two forest rangers, Burnett and Lade, whose first job was to determine how much timber was available on the 657 026 hectares of reserves and to make further 'reservations of well timbered lands where necessary'.

Although the decision to create this department was a major step it fell far short of what was needed. Indeed, it seems to have been little more than a low-level response designed to placate a vocal minority. There was no sign of any forest legislation to support the new department and there was a limit to what could be done by one inspector and two rangers.

Board did the best he could for five years, when he was succeeded by Philip MacMahon in November 1905. Like Board, MacMahon had no formal training in forestry and had previously been a Land Commissioner. His title was changed to Director of Forests and he inherited the two rangers who had worked for Board. It was still a three man department.

MacMahon found that the two rangers had spent much of their time removing areas from the existing timber reserves that were suitable for land settlement, so that they were 'in reality working in the interests of land selection rather than in that of forestry'.

MacMahon was convinced that land must be set apart for forestry and the forests should then be managed. He could see the problems that would occur when a state as big as Queensland was more densely settled. Because timber did not grow across much of the state, settlement in areas that did not produce timber would inevitably have to draw it from areas where it did grow. But there was barely enough growing in those areas to keep pace with local demand.

Within a year of his arrival MacMahon succeeded in more than doubling the staff. He now had four forest inspectors and a junior clerk but even that was not enough, he said, to control 'three and a half million acres of extremely valuable public estate, scattered over an immense area and exposed to attack at many points'.

A year after MacMahon's arrival, in December 1906, the long-awaited legislation was finally passed through Parliament. This was 'An Act to provide for the Reservation, Management and Protection of State Forests and National Parks' and it gave power to reserve crown land as State Forests or National Parks. Their subsequent alienation, that is conversion to private ownership, could be done only by Act of Parliament. It also created a unique situation in Australia in that National Parks came under the control of the forestry department, a situation that was to continue in Queensland for nearly seventy years.

In practice the Act was limited in what it set out to do. It was simply enabling legislation which allowed regulations to be issued under its authority. But the regulations which were to be the substance of the legislation were not gazetted until 1914.

Even if the Act had been more substantial it might not have had the desired effect. Victoria had had a similar Act for the last forty-six years and during that time its forests had been devastated as never

A huge rose gum being felled at Mount Lindsay in 1952.

before, leading the Victorian Premier to say, 'We . . . have acted as goths and vandals to our forests.'

In spite of his small staff, MacMahon worked energetically to put his department on a sound footing. In one year, for example, he was in the field for a total of 145 days, excluding Sundays, he looked at 583 files, read 466 letters and wrote 926. In an age before computerisation it was a daunting workload for a man with only one junior clerk.

He was also very forward-thinking and many procedures that are a part of modern forestry had their germination with him. He saw earlier than most the need to train young people as professional foresters, saying that 'a few of our smart young Queenslanders, selected from country schools and given a special training such as they could obtain in this Branch, would make ideal foresters'.

He also saw the conflict arising from land settlement and, more importantly, the effect that it would have on forestry. In 1907 he wrote: 'The best soils which grow the best softwoods inevitably will have to be surrendered for settlement. It remains to be seen whether the native pine could be established on infertile soils or whether Queensland's future pine supplies will need to come from inferior pine trees adaptable to inferior soils.'

It was a very perceptive view in 1907 and accurately foreshadowed the future policies of the department.

MacMahon's opinions at that time were not universally accepted or even noticed. As an individual he could do little against the ingrained view of the community as a whole and politicians in particular. While the department was edging forward as best it could, it was not having much impact on the world as a whole. On 19 March 1909 *The Times* of London wrote scathingly of what it saw in Australia:

> It is melancholy to have to add that nowhere in the Empire is less practical attention paid to scientific forestry than in Australia, the country of all others where forest administration should be regarded as of the highest importance. The only plea that can be advanced on behalf of the Commonwealth and State Governments is that they are almost overwhelmed by the many urgent questions simultaneously demanding their attention. Yet the need for a careful consideration of forest problems in Australia is very pressing. The wanton sacrifice of timber in every Australian State will certainly bring retribution if it is not checked.

MacMahon died in office in 1910. During his term revenue generated by the department had grown from just over £11 000 to nearly £56 000, and expenditure had increased from six per cent of income to seven per cent. At the time of his death he was responsible for 717 000 hectares of timber reserves, 21 000 hectares of State Forests and 1900 hectares of National Parks.

MacMahon's successor was N. W. Jolly and his arrival in 1911 marked the start of professional forestry in Queensland. Jolly had been South Australia's first Rhodes Scholar and he had studied forestry at Oxford under Sir William Schlich, one of the leading forestry educators in Europe. He was therefore the first trained forester to enter the department since its inception eleven years earlier, and his approach was immediately different.

Almost from the start he identified two important principles. The first was the need to fix an annual cut related to the size of the forest, the population of tree species, their size and growth rate. The second was the need for regeneration so that forest stocks could be built up for future use.

To obtain information that would be necessary to fix the annual cut Jolly established yield plots in which individual trees were measured and their size and growth rate recorded. When there was enough information of this kind it would be possible to make fairly accurate predictions on which the annual cut could be based. Meanwhile he established the cut by observation in each area. It sounds crude but it was at least a start, and Jolly was at that time the only person in Queensland who had the skill to do it, imperfect though it might be.

He also set up silvicultural experiments and nurseries at Atherton in north Queensland and on Fraser Island to determine how forests might best be managed and particularly how they could be regenerated. As with the yield plots, information would come only slowly and until then management consisted of applying whatever seemed to work best by observation.

He also made the first experimental planting of hoop pine, kauri and blackbutt, among other species, as well as planting the first exotic pines and the native cypress. All this work was done before 1916.

Like his predecessors, Jolly continually had to fight to retain the forests from land settlement, but unlike them he did so with the authority of a trained forester. In 1915 he wrote:

> In any case it is not by any means self-evident that land settlement should take precedence over Forestry, for timber is an extremely important national necessity and should not be treated as a product of minor value . . . Also the prevailing idea that timber reserves should be relegated to the back blocks requires to be reviewed, for Forestry is a business which should not be foredoomed to failure. Timber, generally speaking, is less valuable bulk for bulk than most agricultural produce, and more expensive to handle, so that standing trees decrease in value, as the distance from market increases, much more quickly than other crops . . . Certainly, if the Forestry question be looked at from the popular present day point of view, it must be admitted that the returns from virgin forests compare unfavourably with those obtained from the same land under cultivation or even under grass; but this is not a fair criterion, as the virgin forests of Queensland do not yield on an average more than 20 per cent of the yield to be expected when correct forest management has made the land fully productive.

He also pointed out that European countries with their large

FOREVER IS A LONG TIME

Bullock teams with a load of logs at Dayboro, south-west of Caboolture, in the late 1880s.

populations were able to keep between 18 and 32 per cent of their total areas under forest, whereas in Queensland the reservation of only one per cent of the area of the state was thought to be excessive.

There is no doubt that Jolly was the first to turn the department into a truly professional organisation and in doing so he originated much that others were later to claim as their own.

The organisation he inherited was doing its best but it was restricted by lack of funds and by government apathy. By the time he resigned in 1918 he had appointed residential officers at Imbil, Yarraman, Benarkin and Goodnight Scrub. A large nursery had been established at Imbil and six additional officers were now assisting district foresters in other areas. Regulations had been drawn up to control the amount of timber which could be disposed of each year from any State Forest and knowledge of growth rates and silvicultural techniques had increased enormously.

For the first time it could now be said that Queensland had a professional forestry department.

5 The Swain Era

When Edward Harold Fulcher Swain took over as Acting Director of Forests in April 1918 it was the start of an era that was marked by enthusiasm, unfailing energy and constant turbulence. When he arrived he was young and self-opinionated. When he left he was middle-aged, pompous and defeated.

Swain forced himself to excel. With no formal training in forestry he proclaimed his superiority over those like Jolly, who were trained in European forestry and who could not see its limitations in Australia. His staff revered him, but none revered him as much as he did. He had to be the first, the only, the biggest and the best. A colleague said of him that he knew Swain could not have been a Catholic because he was not the Pope. Swain is a legend among foresters even today. He had enormous energy and conviction and he carried all before him. Until the end.

E. H. F. Swain was born in Sydney in 1883, the son of a small merchant. He was educated at Fort Street Model School and left when he was sixteen. His mother wanted him to sit exams to become either a teacher or a railway clerk, but as none of these appealed to Swain he deliberately failed them. He also sat for two other exams, one for entry as a cadet forester and the other held by the chamber of commerce which offered a prize of £10 pounds and the choice of a job with any member of the chamber. He came first in both and decided to become a forester.

THE SWAIN ERA

E.H.F. Swain, photographed shortly after his appointment as Acting Director of Forests in 1918. He remained head of the Queensland Forestry Department until his dismissal in 1932.

He found himself in a 'Civil Service slum' earning £50 a year and if he had been attracted to a bush life he saw little of it. Instead he spent much of his time practising his signature. He also obtained from Oxford University the syllabus of its forestry course but he was not impressed with it because, in the European tradition, it virtually ignored economics. It was, he wrote later, quite feudal.

After a couple of years he escaped from his civil service slum and moved to the forest nursery at Gosford. He then moved on to other forest regions in New South Wales and started to acquire the expertise of a forester. In 1909 he wrote a study of the ecology of the Bellinger River forests which was, he claimed, the first publication of the New South Wales Forestry Department. On the strength of this work he was sent to assess the timber stands of the Bellinger River and this was another first: the first forest assessment in Australia.

In 1910 he was appointed District Forester for the north-west region with his base at Narrabri and he was now a man with a mission. 'I knew that as the first Australian cadet forester, I had to pioneer every step in the development of Australian forestry.'

He stayed at Narrabri for six years and during that time surveyed large areas of the north-western forests and produced maps showing the distribution of forest types. He also developed a system of levelling out the price that sawmillers paid the department for their timber, and this was a matter he pursued with vigour for the rest of his life. It had its origin at Moree in 1912. There he found no less than seven sawmillers along a 65 kilometre stretch of road out of Moree. They all paid the department the same price for their logs, but the sawmiller nearest to Moree could undercut the rest because his transport costs were lower, while those at the end of the line had to steal timber in order to make a profit at the prices that he set.

Swain, with his concern for economics, saw that there was no justice in the department charging the millers the same price because their costs were different, and largely determined by their distance from the market. In Moree he could not reduce these prices because of legislation, so he managed to extract higher prices from those nearest the town.

This was the start of stumpage appraisal, in which the price paid by a sawmill to the department for trees in the forest is worked back from the price of timber in the retail market. From this is deducted the costs that the miller must pay in order to reach that market and the profit he needs in order to stay in business. What is left is the price he pays for the trees.

This had two important effects. It stabilised the retail price because no miller had an advantage because of lower transport costs, and it encouraged sawmillers to work more remote forests. Their profits would not suffer because they would be paying less for the trees. The department's revenue would be less, but the benefit was that the forests could be worked more evenly.

In 1915 Swain took long service leave and at his own expense went to America and spent six months at the forestry school at the University of Montana. When he returned to Australia the following year he published *An Australian Study of American Forestry*. This was the first counter to the European approach to forestry which was so entrenched in Australia.

Americans ran their forests as a business. Forests were a capital investment which were expected to show a return in the future. Controlling costs and using the most efficient management techniques were an inescapable part of this, just as they were in any other business.

The Europeans, on the other hand, showed little concern for the economics of forestry. Because their trees were slow growing the development of forests was measured in hundreds of years and by the time they were harvested nobody had any idea of their capital cost. There was little incentive, and probably little need, for cost-controlled management because the sawmillers were expected to pay whatever was asked of them. And they did because there was no alternative.

Swain maintained his fervour for the American approach for the

A truck load of butter boxes made from veneers leaving South Brisbane in 1932.

rest of his life and he fought many battles over it. While there is no doubt that his conviction was genuine, it also gave him a position from which he could defend his own lack of qualifications as his career developed into prominence.

When Swain returned from America in 1916 he took up the position of District Forest Inspector at Gympie while Jolly was still the head of the department. Typically, Swain found that he was no better off than he had been in New South Wales as 'Queensland too had no forestry'.

Swain was appointed Acting Director when Jolly left and he was confirmed in the position shortly afterwards. At last Swain was in charge of a forestry department and he was like a bee in a bottle. He absolutely throbbed with ideas and his staff loved the challenge that he presented. One said that they were probably given more problems to solve in a month than most would see in a year. They could not always solve the problems, indeed at that time most were incapable of being solved, but they revelled in the stimulation.

Swain claims that when he took over the top job he found some samples of Queensland timbers that Jolly had left on his desk. From these Swain devised a system of anatomically classifying wood that could be applied to all the different types of wood in the world. It was a major achievement, and for those who were unable to recognise it Swain explained it. 'That was the beginning of wood technology in Australia. And in the world!'

Less grandiose but no less important, Swain started to put the emphasis on silviculture, which he saw as the pivot of forestry. He also started to work towards an independent forest department and modern legislation to support it, together with an increase in technical staff and a laboratory to study forest products.

In 1916 the Queensland Government had bought four sawmills as part of a policy of state enterprise and they had been run by the Trade Department. In 1920 they were transferred to the forestry department and they gave Swain valuable information in developing his stumpage appraisal scheme as well as showing how sawmills were an integral part of forestry. Another mill was bought from the Railways Department and a new mill was built at Injune.

That year also saw the planting of the first commercial plantation, the earlier plantings having been experimental. These experimental plantings had revealed an unexpected bonus and that was the triumph of Australian species over exotics. Hoop pine in particular had proved to be very hardy and a good developer in plantations and it was soon to become the focus of the plantation process.

Swain, meanwhile, saw the future very clearly: 'For the most part the forest policy of this State will resolve itself into the provision for Australia as a whole of cabinet timbers and high-grade native pine woods, plus a quota of constructional hardwoods for local and export purposes.'

Plantations were an essential part of this policy. The demand for pine was growing at such a pace that future needs would not be met by pines from native forests. Jolly had recognised this and had started the experimental plantings that paved the way for the first commercial plantings. This much had at least been done by the time Swain took over, but the commercial development was almost entirely his. This development was helped in 1922 when a Deputy Forester called Weatherhead invented a method of growing and planting pine seedlings.

Pines had been planted open root into prepared ground and this had to be done in the winter. The next growing season, however, produced a crop of weeds that were often so vigorous that they would stifle the growth of the tree seedlings when they most needed it.

Weatherhead's invention was delightfully simple. It consisted of a sheet of galvanised iron bent to make a cylindrical tube and with

THE SWAIN ERA

A consignment of cabinet-wood stumps awaiting shipment to the USA in 1929 for use as veneers.

the edges overlapping by about a centimetre. The edges were held closed by a circular band of tin that was slipped over the tube and which restrained the natural spring of the iron and thus tightened the tube. The tubes were placed in a concrete bed and filled with compost. The seed of the tree was planted in the tube and allowed to grow until ready for planting out.

The site was prepared by clearing and burning the debris in the middle of summer and as soon as this was done the tubes were taken there for planting. A hole was made with a mattock and the tube, containing the seedling, was dropped in. The band was removed and this allowed the tube to expand slightly. Earth was packed loosely around it and the tube was then withdrawn, leaving the seedling well established in its native soil and in its permanent location. By the time the weeds grew the seedling was able to compete.

This invention, the Weatherhead tube, led to much more efficient and more certain planting of plantation stock and it is still in use today.

In October 1924 Swain realised his ambition of having an independent forestry department, although it was not entirely as he would have wished. The department was no longer a branch of the Lands Department with its own director. Instead a new authority was formed called the Provisional Forestry Board which was answerable directly to the Minister for Lands. Desirable though this was, the flaw for Swain was that although he was to be head of the Board there would be two other members, which meant it was not his alone. At the same time the regulations that had been made in 1914 were revoked and new ones came into effect in November.

The new department consisted of three branches, as proposed by Swain. These were harvesting and marketing; administration and accounts; and working plans, silviculture and surveys. The state was divided into six regions and in each of these the marketing of the old crop was kept separate from the production of the new. The branches operated independently in each region and were coordinated by the Board in Brisbane. This principle has survived until the present day, although in many different forms.

Soon after it was formed the Board investigated its forest resources and the rate at which it was being cut against the need to supply stock in the future. The Board forecast that the pine forests would be cut out in ten to fifteen years and so it decided on an annual programme of naturally regenerating 6000 hectares of native forest and the planting of 2000 hectares of plantations.

To support this an experimental station was set up at Beerwah to examine the use of exotic pines from America in the wallum land that ran along the coast north from the New South Wales border. Although extensive, this land was regarded as unproductive and was therefore free of pressure for clearing and settlement. It was ideal as a site for large plantations provided a species could be found that would thrive on it.

The dilemma facing the Board, however, was that pine from the native forests would cut out long before plantation pine was ready for harvesting. They had a stock of native forest pine that would last no more than fifteen years, or until about 1940, and plantations that could not be harvested until the early 1950s. In between would be a gap of a decade or more during which there would be no pine available if it continued to be cut at its present rate.

The answer, obviously, was to reduce the cut. In 1926 the Board announced that the cut of hoop and bunya pine would be reduced by about three million super feet each year until 1952, and they recognised that this would mean at least one sawmill would go out of business each year.

Another way to cope with this reduction was to divert some of the uses of pine to other species. Although the state had a remarkable number of species available, sawmillers and end users had in practice used only a few of them. They knew them well and as there

THE SWAIN ERA

had been a seemingly endless supply there had been no need to go beyond them.

Swain drew attention to the range of woods that were available in his book *The Timber and Forest Products of Queensland* which was published in 1928. This book, written mostly by his staff, describes over 200 species in great detail and is still a standard work. He also set up the Forest Products Laboratory and staffed it with an excellent team of scientists. The laboratory, the first of its kind in Australia, investigated the suitability of timbers for specific purposes by carrying out tests and long term trials which were beyond the scope of industry. It also offered advice to industry and encouraged architects and other users to consider species that they had not used previously but which the laboratory could show were suitable for their requirements.

Apart from becoming independent, the department had grown considerably since Swain had taken over. Some of this growth had been in specialised areas, such as the laboratory, but the increase in silvicultural work had also increased the need for field staff. Most of these were people who had worked for the Lands Department but they were later augmented by returned servicemen. The Federal Government offered a subsidy of 25 per cent of the wages of a returned serviceman and Swain later described this as his first break:

> I collected returned soldiers who had surveying experience, deployed them in Forest Assessment survey camps, and drilled them in such surveys . . . Over the years, before aerial surveys, these soldier parties ranged from north and south, mapping and tallying, using for contouring the Bonner Abney cord tape which I had brought from USA. It was the toughest of pioneering, manfully done, and deserving decorations. Their forest mapping has received world encomium. It included subdivision of the forests into hundred acre compartments and the economic design of forest roading . . . This compartmentation became the framework of controlling forest management and accounting, providing a basic order. Except for the north-west of New South Wales, where I initiated it, it had never been done in Australia before.

All this work needed financing, however, and here he ran into problems. In the past the department had usually shown a surplus of income over expenditure and this had been handed over to the government. Indeed, expenditure was usually modest because very little had been done in terms of forest development, hence the financial surplus.

Swain's plans, however, were of a quite different nature. It was, for example, sound financial sense for the department to build roads

GROWING UP

Taking plants to a new nursery site at Brooloo in 1932.

through its forests. Because the stumpage appraisal system meant that the sawmiller's transport costs reduced the price he paid for the trees, good roads would reduce those costs and the cost of building them would eventually be recovered by way of increased stumpage. But they had to be built in the first place.

Similarly, the cost of developing plantations was considerable, especially in their early years, and the income from them would not arrive until thirty or forty years later.

Politicians had difficulty understanding this. They recognised the need for capital works such as railways because the benefits were felt by their voters relatively quickly. But the need to protect against a shortage of timber that would not be felt for another twenty years seemed far less pressing.

In the end, and after much criticism, the government agreed to finance capital works in the forests by means of a loan fund and until the Depression this averaged about £35 000 a year. The use of loan funds for such work is unremarkable now, but it created a great deal of heat at that time.

Swain established the practice of paying the costs of marketing available timber from the income it produced, and using the loan funds to establish the plantations. At the same time he set about acquiring land for forestry purposes, and particularly for plantations.

In 1918 Acting Premier Theodore, on Swain's advice, had said that over two million hectares would be allotted to state forests and 400 000 to timber reserves, but this was not carried out. Most of the land that *was* allocated went into timber reserves, which could be revoked at any time.

The nature of the land was also a problem. There was still a view that good land should be used for settlement and agriculture and that forestry should therefore be confined to land that was remote or otherwise unsuitable for those purposes. In spite of his efforts, Swain was unable to convince politicians that forestry required good land that was close to markets. Even when they agreed that forestry was a worthwhile use of land, they were unable to accept that it should be done on land that was suitable for other uses. Even the timber reserves came under criticism because unless they were actively being logged most people could see no purpose in them.

Swain, like his predecessors, was under constant pressure to release existing forest land for settlement, especially in northern Queensland where the native forests stood in the way of further settlement. Many plans for settlement were ill-conceived. They did not consider the suitability of the land for agriculture, the cost of marketing the produce in distant centres or the problems of over-production should agriculture prove successful. But the demand for land settlement in north Queensland in the 1920s was almost insatiable.

It was also in line with government thinking. North Queensland had to be settled in order to dissuade Asian countries from trying to take it over, or so they thought. And perhaps with some reason. Cairns is about as far from Brisbane as is Melbourne and in the 1920s, without modern transport, that was a very long way indeed. Yet north Queensland had a climate that was similar to countries of southern Asia and could be put to good use by people from there who would have revelled in its space compared to the population densities they were used to.

Whatever the merits of this argument, the fact remains that settling north Queensland was seen as an urgent need and requests to throw land open for settlement were usually well supported by politicians and by the Lands Department. The only opposition came from the forestry department. Swain's plan was to develop the cabinet-wood industry in North Queensland and the thought of giving the forests away to people who would loot its valuable timbers appalled him.

A parliamentary party committee had examined the conflict between land being used for forestry and the need for settlement and had recommended that 160 000 hectares should be reserved for softwood plantations in southern Queensland. It also recommended the setting up of a Forest Boundaries Committee which would hear complaints and advise the government of competing claims for land use and especially in allocating the land required for plantations.

In the south this was largely uncontroversial and was carried out with a fair degree of consensus. But it was quite different in the north. There, the conflict had become so hostile and the positions so entrenched that in May 1931 the government set up a Royal Commission to examine the whole question of forest boundaries in the north. Among other matters, the Commission was asked to inquire into:

- The situation and areas of virgin land situated north of Ingham that are suitable for closer settlement.
- In relation to Forestry Administration:
  The land which should be permanently reserved for State Forest purposes.
  The land which should be temporarily withheld from settlement pending the disposal of timber thereon and the best means of marketing such timber.
  The lands which should be immediately made available for settlement.
- Generally a definite programme of land settlement and forestry activities in the tropical north for the ensuing decade.

To most observers of Queensland forestry, then and later, the Royal Commission was a 'stacked deck' designed to throw forested land open for settlement in a way that the government was unwilling to do by legislation. It was seen as a means of arriving at a predetermined solution with a minimum of political fall-out, and if so it would not have been the first Royal Commission to be set up for that purpose.

The choice of commissioners certainly seems to support that view. They consisted of William Labatt Payne, Chairman of the Land Administration Board; Winton Woodfield Campbell, staff surveyor with the Department of Public Lands at Innisfail; and John Grainger

Atherton, 'formerly of Chillagoe and Cairns'. None came from the forestry department, or could be expected to represent its interests.

For Swain it was a call to arms. With the Commission intending to hold its first public hearing at Ingham on 9 June, Swain headed north to argue his case in public before they arrived. The Commission telephoned the Premier to have him recalled, but by this time Swain had joined up with his Minister and so was politically immune.

Later, Swain wrote:

> The Commission left me to be its very last witness because it had

A horse team hauling logs from a forest in the Brisbane Valley in 1925.

lacked confidence and thought to better verse itself first. But this enabled me to complete my 200 page report on the development of the north and I had a hundred copies mimeographed. The first copy reached me only five minutes before I was called.

The Commission ordered the press out, saying that it would hear me in camera. I bowed politely; and for two days I read my report including unwelcome excerpts from the files of the Chairman's own Lands Department.

The Commission went pink, green and white!

Finishing, I thanked the Commission for its patient hearing, offered to supply any further information it desired, bowed myself out, returned to my office and said: 'Right! let them go'—dozens of copies of my evidence in camera to the press of Queensland!

Swain was far too smug, and far too soon. If he thought he had won the battle, the report of the Commission showed that he had not won the war. This report has often been taken as an example of the ignorance the community as a whole showed towards forestry and it has been selectively quoted to support that assertion. But a closer reading of the report shows it to be more objective than its detractors suggest and it is difficult to avoid the conclusion that Swain went too far in his dealings with the Commission and his arrogance led him to make statements which could not be justified.

The report commented on Swain's trip north ahead of the Commission and after acquitting his party of any improper motive said: 'We think also that due allowance must be made for the contentious atmosphere with which Forestry officers, by their contentious and aggressive propaganda, have succeeded in surrounding themselves . . . Because of this, Forestry in Queensland is fast becoming more a problem in psychology than a problem in economics.'

Commenting on the paper Swain had presented to the Commission, the report said: 'Brilliantly worded with typically audacious and reckless propaganda, fundamentally wrong in argument, grossly inaccurate in figures and in narrative, it reads in parts more like imaginative fiction, than sworn evidence . . . More eloquently than anything we might say it illustrates the impossible contentions of the Queensland Forestry Department.'

The report went on to examine those contentions. If the department had its way, it said, not another acre of crown land would be opened for settlement in North Queensland. But with a growing population the need for food was paramount, while the need for timber was not. In any case there was timber enough. And it said, in a piece usually quoted out of context by those supporting the forestry cause: 'The productive wealth of the country at present

A distinctive Linn tractor with a full load at Villeneuve in 1929.

suffers from the fact that there are too many, rather than too few trees.' To a forester, of course, that was outrageous. There can never be such a thing as too many trees.

The Commission also made the point that resources are often replaced by others. Oil was replacing coal and it suggested (in 1931) that atomic energy might in turn replace oil. It questioned the wisdom of growing timber on the assumption that there would never be a substitute when doing so removed the land for any other purpose.

The report then examined in detail some of the statements made by Swain and his officers and, in measured terms, blew them out of the water.

Swain had said that soil deteriorated when the forest canopy was removed. But, the report said, the production of sugar cane had increased from 1.19 tons per acre in 1900 to 2.41 tons an acre in 1929, while the production of maize had remained steady after thirty years of cropping. The examples Swain had quoted, bananas and pineapples, had fluctuated because of the market, not because the soil had deteriorated.

Swain had also said that timber was the economic mainstay of the Far North but, the report said, sawn and log timber provided only 8 per cent of the annual production of the region. It went on:

Another illustration of false propaganda is the claim made by the Deputy Forester at Atherton that 6,000 men are directly employed in North Queensland in the harvesting and marketing of timber. Six thousand men at four pounds a week would receive 1,248,000 pounds in wages per annum. The total value of log and sawn timber produced in the region under reference during the year 1929–1930 was, however, 465,704 pounds. The falseness of the statement is manifest.

The final blow was direct and to the point: 'A spirit of enthusiasm for Forestry permeates the State Forests Administration. No one can fail to note or to be impressed by its presence. Well-directed it would be a most admirable quality. Ill-directed, reckless and unfettered, it is becoming a real danger.'

The Commission listed twelve recommendations which would reduce public expense of forestry administration, increase the area of forest reserves and National Parks by about one and a quarter million hectares, and make available for settlement in the near future an area of about 30 000 hectares.

Swain had grossly overplayed his hand and had been discredited as a result. In 1907 MacMahon had recognised that some of the best forest soil would have to be surrendered for settlement, but to Swain this was unthinkable. Not one acre. Caught up in his own ideology, he had been unable to tolerate any compromise or accept that the community might reasonably argue that it needed forest land for other purposes. Balancing these needs did not always mean unlimited reckless clearing.

The Premier proposed to send the report to the Attorney General and to prosecute Swain for perjury. This was not followed up, according to Swain, because the Auditor General confirmed Swain's figures were accurate. If so, that information was not tabled in Parliament. Only the report was, and that was enough.

There was a state election almost immediately afterwards and then Swain went on holiday: 'I returned to my home on the Friday ready for work on the Monday. There was a knock at the door, and my weeping secretary with a large official envelope in his hand. It was my dismissal, without notice or explanation . . .'

It was the end of Swain's career in Queensland but not the end of his career in forestry. In 1935 he was appointed Commissioner to the New South Wales Forestry Commission, where he stayed until 1948, and in 1950 he went to Ethiopia as a forestry consultant on behalf of the United Nations. He died in Brisbane in 1970.

6 The War and the Aftermath

Swain's sacking in 1932, dramatic though it was, failed to rate a mention in the department's next annual report. It was, of course, so well known that it was perhaps considered unnecessary. On the other hand, it almost seems as if the administration was anxious to purge itself of all memory of him.

His place was taken by V. A. Grenning, and he had no doubt of the huge task ahead of him. Swain might have been disliked by many outside the department, but he had generated a great deal of enthusiasm and loyalty within. Grenning's immediate task was to hold the department together and to keep it functioning.

He had the advantage of already being a member of the department that he now took over. He had joined it in 1922 after completing his studies at Oxford as Queensland's Rhodes Scholar of 1919. On his way back to Australia he spent time in India and the Pacific gaining first-hand knowledge of tropical conditions.

The Provisional Forestry Board was now disbanded and a Sub-Department of Forestry was formed as part of the Department of Public Lands, with Grenning as Director of Forests. And, in spite of the upheaval, the department got on with its job.

A lot of work was being done on the use of Queensland timbers and particularly in the technology of seasoning, and more was being learnt about growing plantations. A new method was introduced which aimed at keeping the planted area completely clean during

Building a track to Palm Valley in the National Park at Tamborine in 1938.

the first year. This was done in three stages. The first consisted of chopping out all the inkweed when it was between 150 and 300 millimetres high and this was followed by the second stage in which all weeds were completely removed by hand pulling or grubbing. The third stage was usually not necessary but it consisted of further hand pulling or digging out of woody weeds or late germinating softweeds. By this time all plantation weeds had been classified and a treatment had been laid down for each. This method of keeping

THE WAR AND THE AFTERMATH

the ground clear proved very successful in its first year. Bad conditions in 1932 would, under the previous method, have seen significant losses which were very much less with this new method.

Another significant step forward was in the preservation of hoop pine seed. In order to extend the area of plantations there needed to be a steady supply of seeds to be germinated and grown as seedlings in the nurseries. No seeds, no trees.

The difficulty was that hoop pine does not produce seeds every year. The fall of seed might take place every three or four years or even longer and is impossible to predict. Although the seed fall, when it happened, was abundant the seeds deteriorated after about twelve months and became useless. In 1928 a quantity of seed had been put into cold storage at about four degrees centigrade and four years later was tested for germination. The seeds were found to have retained their original germination capacity and were as good as when they had been collected. This method is still used today and the department maintains a large stock of plantation seed in cold storage.

Surprisingly, the Depression seems to have had little effect on the department although it did affect its own sawmills. These were now losing about £4000 a year and as there seemed no likelihood of improvement they were sold to private buyers. In the forest, though, business was getting better all the time. In 1933 the officer in charge of the Mary Valley District said that there was not one genuine timber worker unemployed in the district.

The following year was even better. The sale of logs in North Queensland was much higher than it had ever been and the revenue of the department reached an all-time high of over £600 000. Much of this came from the sale of hoop and bunya pine. This was not a reflection of increased building activity, which would normally have been the case, but the decline in supplies from private forests which were now virtually exhausted.

It was also a sign of the increase in mechanical logging. Caterpillar tractors were now widely used for snigging logs out of the forest and trucks were used to haul them to the mills. These machines were much more efficient than bullocks and horses. They did not depend on weather to produce their feed but they did require better roads. This high level of cutting, though, began to sound alarm bells. The concept of sustained yield was now well established, but it was not being maintained. Sustained yield is the practice of cutting only as much timber from a forest in a year as it has grown. If the forest grows one million cubic metres of new wood each year, then one million cubic metres of old wood can be cut from the forest each

Loading a truck with tubed hoop pine seedlings at the Imbil nursery in 1940.

year. In this way the forest is never diminished and retains its capacity to regenerate and produce more timber for the future.

Sustained yield was, and still is, one of the basics of modern forestry. When successfully applied the forest can produce timber in perpetuity, which is the fundamental aim of forestry, because the amount of the wood in the forest never falls below its original volume provided adjustments are made for changes in growth rate as a result of climatic changes or natural events such as cyclones.

As a result of the current high cutting, however, the annual cut now exceeded annual growth and that meant that the forests were actually being diminished. On the overall scale, this was partly the result of clearing. The Eungella State Forest, for example, had recently lost 6500 hectares for selection as dairy farms and this meant that the total growth in the state's forests was reduced accordingly. Unfortunately the demand for timber was not reduced.

There was also, for the first time, an awareness that forests had other uses besides producing timber. Grenning wrote: 'As the years pass and the natural vegetation is destroyed on settled lands, more and more people are finding in the forest areas a source of enjoyment and health. The forests are the home of many strange and beautiful plants—too tender to survive outside their limits—whilst birds and animals which otherwise would become extinct are given a chance to survive.'

The forestry department was still responsible for National Parks and in 1936, after Grenning had visited America and Canada to study practices there, a programme of development and protection was introduced along American lines. The stated principle was 'the preservation of the scenic, educational and recreational values of the areas set aside'. At that time the conservation of nature was not part of the aim other than as might be necessary to support the overall principle.

The Secretary of the Department, C. J. Trist, took over the administration of National Parks and oversaw the developmental work. This consisted of establishing walking tracks and providing other facilities for visitors. Even so, the department faced opposition in its handling of National Parks. It was criticised for not allowing logging on the one hand, while others saw National Parks as ripe for commercial development and wanted to turn them into resorts.

Parks were either small areas around special scenic spots, and there were many of these, or they were areas of natural scenic attraction that had previously been logged. This meant that many areas were overlooked, such as swamps and mangroves, which had value specifically in terms of nature conservation rather than as beauty spots.

Meanwhile the cut from State Forests continued to increase. The total cut in 1936–37 was 162 million super feet, the greatest ever. On the other hand, the results of tending plantations were becoming apparent. One hoop pine plantation which was now twelve years old was carrying 12 000 super feet per acre, which was larger than the average stand of hoop in native forests.

The promise that this offered was in the future, however, and it did nothing to balance the overcutting that was going on now. In an attempt to bring this under control an Act was passed in 1936 which required all sawmills to be licensed and, it was hoped, their number reduced. By the following year 600 mills had been given licences, a huge number to be drawing on the state's limited timber resources.

By the late 1930s the department benefited from an abundant labour supply as a result of the Depression and Grenning put it to good use, just as Swain had with the returned servicemen. The total number of staff more than doubled in two years as nearly a thousand men and youths moved from the cities to the bush on full-time relief work. They were used on labour intensive activities such as building fire lines and roads and carrying out the new regime in the plantations. It was to be short-lived however as the numbers diminished rapidly with the outbreak of war.

Building a forestry road in the Kalpowar district in 1939.

It was clear right from the start that the war would put an enormous strain on forests all over the country and the Federal Minister of Supply and Development immediately called a conference in Melbourne to consider the anticipated reduction in imports of softwoods from overseas.

As a result of previous care Queensland found itself in a relatively fortunate position as it was able to meet its own needs and was also exporting to other states. Because of the concern about overcutting there had been a move to reduce the cut from native forests of hoop, bunya and kauri pine. This was now abandoned and the cut was actually increased to meet the needs elsewhere.

Hoop pine became even more important than before. It was used to make paper when imported pulp became unavailable and the whole of Australia's butter export was packed in cases made of hoop

THE WAR AND THE AFTERMATH

pine from Queensland. During the first year of the war a record 1215 hectares were planted with hoop.

In 1941 the Conservator of Forests in Western Australia, Stephen Kessell, was appointed Controller of Timber for the Department of Supply and Development. His job was to maintain the supply of timber for military purposes but in practice he controlled the supply of timber for any purpose throughout the country. It was impossible for anybody to receive timber from anywhere without Kessell's approval.

Kessell delegated the Queensland department to act on his behalf in the state and to make the best use of its timber and sawmills. Much of the logging plant of contractors in north Queensland was requisitioned for defence work and the older equipment they had to use was limited by the lack of spare parts. In spite of this the cut of mill logs was greater than ever.

Within the department staff numbers fell at an alarming rate. In 1938–39 total staff was just under 2000, but by 1942 it had fallen to 719. This meant that no further tree planting could be carried out and there were barely enough people to carry out basic maintenance. As if to make matters worse, that year saw the first fall of hoop pine seed since 1936. It produced over eighteen tonnes.

As the war progressed the department had barely enough people for fire protection and even though timber production was, incredibly, higher than any pre-war year it was still not enough to meet demand. The department reported: 'The average quantity of logs processed by the Queensland mills during the three years immediately preceding the war was 282 million super feet. Yet when the innumerable difficulties of man-power and plant experienced by the logging and milling industries are taken into consideration, it is surprising to find that the industry was able to exceed the pre-war figure by no less than 30 million superficial feet per annum, and to log and manufacture an average quantity of logs during the six war years of 312 million super feet.'

It was an almost unbelievable achievement, but it had been at enormous cost to the forests. With little supply of pine from private forests, the department's forests had been hacked mercilessly. The State Forests supplied 68 per cent of the entire cut during the war and 94 per cent of pine.

The end of the war was not the end of the problem. Attention now turned to supplying post-war needs, such as housing, that had been neglected during the war. Grenning and his staff had drawn up a post-war forestry programme which was implemented as soon as the war was over. However, this plan called for a staff of over 1800

GROWING UP

Married quarters at Beerburrum in 1950.

people and they had nothing like that. Those they had were allocated to the first priority, which was planting.

The shortage of manpower became one of the biggest obstacles to the reconstruction programme. Even as men became available most had little enthusiasm for forestry work. By the end of 1947 the department had taken on 1100 men, but only a hundred had stayed. This shortage led to increased mechanisation, but there was no real alternative to manpower for many of the jobs which needed doing in the forests. So in 1948 Grenning put in a request for 250 European refugees to help with the essential work in pine plantations.

These men, known at all levels as Balts because most came from Baltic countries, became an essential part of the reconstruction programme and it could not have been carried out without them. They reached Australia as part of a resettlement scheme and they arrived with few possessions, little money and in most cases no English. Theoretically they had to do whatever work was assigned to them for two years, after which they were free to do what they wished. In practice there was little to stop them drifting off at any time, although very few did. Many went to work on the Snowy Mountains Hydro-Electric Scheme while others found themselves in the forests of Queensland.

They had no families in Australia. Wherever they worked, that was now their home. And they worked very well. Most were young and soon grew fit even if they had not been previously. They were cheerful, they lived as hard as necessary and played equally hard

The dining room at Toolara nursery in 1959.

whenever they had the chance. They learnt English, they started to save a little money, and built a new life for themselves in a strange land.

By 1949 there were 400 Balts in Queensland's forests and they lived in groups of varying sizes in places which most born Australians would have found remote and demanding. The first arrivals lived in tents in the forests but as the numbers increased it was obvious that more permanent accommodation was needed. This was hardly lavish. In 1948 the department started to build barracks to replace the tents. The barracks were timber buildings consisting of four single rooms with a communal eating room at one end. Cooking facilities and showers were outside.

A few barracks were built for married couples and these were a little different. The walls were timbered to waist height and the roof was a double layer of canvas over a long ridge pole. The accommodation consisted of three rooms: a bedroom at either end and an eating room in the middle. Immediately outside the door was a cookhouse made of galvanised iron.

By 1950 the department had built eighty-five barracks and these were the only home that most Balts had while they were engaged in forestry work. Some are still in use as temporary accommodation.

Although the Balts made it possible for plantation work to go ahead, cutting of mature timber went on unabated. There was another record cut in 1951–52 and by the middle of the year there were 1284 sawmills licenced in the State, an increase of 121 in the last twelve months. Grenning wrote: 'Permanence in the sawmilling industry can be achieved only by restricting the cut of sawmills to the capacity of the forests to produce timber. Over most of Queensland the capacity of licenses already granted greatly exceeds the growth of the forests.'

Many new sawmills had been granted licences because they had access to nearby private timber. But when that had been cut out they looked to the State Forests to provide timber to keep the mill operating. Forestry, though, was about maintaining forests to meet a demand in perpetuity, not about keeping other people in business by providing raw material on demand.

In 1954 Grenning drew attention to several points, most of which had been made before but which were increasingly important now. The first was that there had been a big increase in cut mill timber to meet the post-war demand and that most of this increase was coming from hardwoods on private land. The second was a marked fall in the cut of pine from crown land. This confirmed the prediction that had been made in Swain's time, that pine would be cut out of native forests before the plantations were ready for harvesting. Pine had already been cut out of private forests and now the supply from State Forests was diminishing as well.

Although the increased cut was incapable of supporting all the sawmills, it was also barely able to meet public demand. The timber producing areas, which the department had in the past fought vigorously to preserve, were confined to a relatively small area in south-east Queensland and to the Cairns district. These were not able to meet the demand for reconstruction at this time and Grenning continually had to explain that timber could not be produced overnight.

Soon even the plantation programme had to be reduced because of the lack of government funds. Staff was reduced until it was nearly at war-time levels and once again there were barely enough people to ensure adequate fire protection. The same financial restrictions also made mortgages difficult to obtain and the housing market went into decline, with a corresponding reduction in the demand for sawn timber.

In December 1957 there was yet another change to forestry administration, although this was more welcome than most changes in the past. After twenty-six years as a sub-department of Public Lands, forestry was again made an independent department and it was to report directly to the Minister for Agriculture and Forestry instead of the Minister for Lands.

THE WAR AND THE AFTERMATH

Important though this was, in practice the department had enjoyed de facto independence for some time. The skills of forestry had now become more technical and the department was responsible for putting an essential product on the market. The department had little in common with the Lands Department and, sensibly, that department had left it largely to its own devices.

At almost the same time Parliament also passed the *Lands Acts and Other Acts Amendment Act of 1957*. This clumsily titled legislation at first seemed to have little to do with forestry, but in practice it had a great deal to do with it. Under this Act holders of leasehold rural property could apply to have their titles converted to freehold. The department was faced with the task of valuing the timber on those properties so that it could be brought into the cost of conversion. It was a process that was to involve members of the department for years.

Two years later the *Forestry Act of 1959* finally gave legislative support to the forestry department and gave the title of Conservator to its head.

This Act said that the 'cardinal principle to be observed in the administration of State Forests' was 'the permanent reservation of such areas for the purpose of producing timber and associated products in perpetuity'.

Some writers have since complained that this was a utilitarian approach which saw forests entirely in terms of wood production. Yet that is primarily what forestry is about.

By the end of the 1950s the department was described as having a 'level of excellence unrivalled in Australia at that time'. It had led the way in technical advances of many kinds and now had considerable expertise in plantation work including seed selection, tree breeding, site requirements, and yield regulation. In the native forests there was a growing knowledge of silviculture, especially in tropical areas, together with highly sophisticated inventory survey systems.

The *Forestry Act* also gave the department slightly more secure tenure over the land it managed. The shortage of timber during the war and the years following had finally made politicians and the public generally more aware of the importance of timber and the continuing need for it. When the coastal ranges were covered with trees it was difficult to see any problem arising in the future. But when people could not build a house or pack produce in a box because of the shortage of timber there was a growing recognition of its importance.

Some people, especially in the north, still saw forests as an

GROWING UP

obstruction to dairy farms, but elsewhere there was a willingness to recognise that forests were an essential part of the state and that they had to be preserved.

This awareness had been a long time coming.

Victor Fedorniak
Cossack–Australian

Those who came from Europe after the Second World War and worked in forestry were known as Balts because most came from the Baltic countries. But Victor Fedorniak was not from the Baltic, he was from the Ukraine, which was one of the republics of the Soviet Union, and that made him a Ukrainian Cossack.

He was born in 1928 in Boryslaw, a small town in the west of the republic. He was educated there and in the nearby city of Drohobycz and it was there, during the war, that he made a decision that was to change the direction of his life.

'Young people of my age were quite heavily involved in underground work. We distributed pamphlets against the Germans and against the Russians. And we used to do a bit of smuggling, guns, things like that which the adults weren't game to handle. When the Russians started to come closer in 1944 we decided it wasn't safe to stick around so we slowly moved west.'

At the age of sixteen Victor Fedorniak left his family and moved to Regengburg in south-west Germany, where there was already a significant Ukrainian community. They were waiting for the war to end so that they could return to the Ukraine and create a new State. However when the war did end the Ukraine remained part of the Soviet Union and there was no way back.

Helped by the United Nations, Victor Fedorniak completed his education at Regengburg in 1948 and worked for a few months in a sanatorium. As the German forces returned to civilian life they were given preference in work and Victor was stood down. He worked for a few months on local farms but there was clearly no future in Germany for a young Ukrainian.

People like him were now encouraged to leave Germany and Victor Fedorniak chose to come to Australia. A couple of his friends had already left for England and others had gone to Belgium, but most went into coal mines and that had no appeal to Victor. Not that he knew much about Australia. 'I knew it was a big country and raised a lot of sheep and was a British protectorate or colony. But it seemed like a country where you could do something.'

After a thirty day voyage on the SS *Oxfordshire*, which seemed like a holiday, Victor Fedorniak arrived at Newcastle on 13 March 1950. He had one English pound, sent to him by one of his friends, a few books and some modest personal possessions. He had barely enough English to keep out of trouble.

He was taken from Newcastle to a refugee camp at Greta, where there was a wide range of nationalities. The refugees were interviewed for several weeks before being allocated to work. They had no choice in this and were likely to be sent to work on roads, in sawmills or in forests. Victor Fedorniak was one of a group of refugees detailed to work at the State Forest at Kenilworth. It meant little to him. 'The war was hell and after that you could accept anything.'

When they arrived at Kenilworth they lived in tents and had to look after themselves. Their work consisted of planting, cutting timber, and clearing undergrowth and weeds. They used axes, cross-cut saws, brush hooks and lots of muscle.

Living was fairly cheap for those who were not reckless. They formed a soccer team and their games, together with Saturday night at the pictures, were their main source of entertainment. Another 'entertainment' was trying to learn English, which Victor did at evening classes at the local school. His foreman also helped him in the evenings and Victor practised by reading comics. The pictures helped to explain the words.

His relationship with the Australian staff was good most of the time. 'There was always the odd one who thought you were there as a slave. I never had any grudges against any of them. I had a few arguments, but the majority were quite good. I can't complain. The majority of families in Kenilworth accepted us and tried to make us feel at home.'

Although he had no intention of staying at Kenilworth any longer than he had to, that changed when he met Fay. By then he had managed to buy a motorbike, he had £60 in the bank, and so in 1953 they were married.

They moved into married quarters, which meant a different tent, and Victor started to grow bananas in his spare time. Even when Fay was six months pregnant they still went to the pictures on the motorbike, but with the money he earned from the bananas he was able to buy a ute.

Three years later, in 1956, the man who had been running the nursery died and his job became vacant. Victor had been thinking of leaving forestry but decided to apply for the job. 'Probably with a bit of education I had a chance of getting it even though my English was not the best. I don't know why they gave me the job to be honest. Perhaps because they knew the local family I had married into, or maybe they thought I would be good at it.'

A few months later the Fedorniaks left their tent. By this time Victor had lived in a tent for six years and with Fay for over three of them. Fay's uncle had a farm with a house on it and he suggested they might move in and Victor could work in his spare time instead of paying rent. After they had done this for a few years the uncle suggested they buy the place from him on interest free terms.

'I agreed to that. I had three boys by then. They come quickly. I bought that farm. I used to get up early and get the cows in and get started. Fay and the boys finished the milking and I went to the nursery for the day's work. It took me twelve years to pay it off. Then we owned it.'

Meanwhile he taught himself how to run a nursery. He did a correspondence course with the Queensland Agricultural College at Gatton which gave him an insight into propagation and soil types, but he mostly taught himself by reading. But he refused to consider himself an expert, even just before his retirement in 1993 as senior overseer of the nursery. 'I can say I am pretty conversant with growing hoop pine. But the other stuff, if you are not doing it you forget those things. I am just average. My knowledge is not expert knowledge.'

When he took over the nursery it was producing about 60 000 plants a year. The nursery was extended in the 1970s to meet the demand for new plantations and is now capable of producing half a million plants annually. They produce only as many as required, however, but in his last year they potted nearly 230 000 plants. 'The first crop I grew went out in 1957 and I sometimes go and look at them. They are pretty large trees now.'

Victor Fedorniak arrived in Kenilworth in 1950 as a refugee and he retired there in 1993. During that time he served three years as chairman of the local P & C and was actively involved with the Chamber of Commerce, as well as many sporting bodies.

His most treasured memory is of an event that took place in 1978. He had not seen or heard of his mother since leaving the Ukraine in 1944. He had tried to find her but she had remarried after the death of his father and her name had changed. But she was also trying to find him. There was no Red Cross in the Soviet Union but she had a friend in Poland. She wrote to her, and the friend asked the Polish Red Cross to help. In 1978 Victor received a letter from the Red Cross containing his mother's address. They corresponded regularly until she died in 1988 at the age of eighty-four, but they never met again.

And retirement? 'I am going to miss the nursery. I can feel it now when I wake up at night and start thinking. What am I going to do when I stop working? Nobody to argue with!'

7 Changing Needs, Changing Views

The period from the 1960s to the present saw more changes and pressures on the forestry department than ever before. Some were internal, the result of vast changes in technology which put heavy demands on those involved and the need for more efficient internal administration. Others were external and were a result of a dramatic change in the public's perception of what it wanted from its forests. These changes put a heavy strain on the department. Everybody involved with forestry had to change, and many found it a difficult process.

There is no doubt that change was needed. The department was now mature and a leader in forest technology. It knew a great deal about what it was doing and it did it successfully. Or it did by its own terms, which were those of professional foresters. But the department had become conservative, inward-looking and slightly arrogant. Some would say more than slightly arrogant. Because it was so technically efficient and well trained it had forgotten that it was running the state's forests in trust for the public. Many in the department had come to think that the forests belonged to them, that they knew what was right for the forests and that the public should not question, or even show much interest in, what they were doing.

The department had, with the passing of time, become more remote from the public it served. Most of its members had passed

A tractor clears an area of wallum as part of the site preparation for an exotic pine plantation near Beerburrum in 1970.

through identical training courses and they had inherited the traditions of forestry without questioning them. They were the professionals and forestry matters could safely be left to them. It was so self-evident to them that they failed to recognise that there could be a different view.

In some respects they were right. Forestry had now become a highly skilled technology and one not readily understood by the public. It had many facets and each was a speciality in its own right, so much so that 'general purpose' foresters, especially those in the field, often had little knowledge of these specialities. Genetics, soil chemistry, surveying, timber research, biology—all were now an essential part of forestry. Forestry had become so technical that most foresters could not grasp every aspect of it.

If the public understood even less, that was largely because foresters had made practically no attempt to explain forestry to them. When they did, forestry was already under pressure and it was too late. Times had changed. The public had formed views of its own that challenged the accepted wisdom of foresters and, for the first time, foresters had to justify what they were doing.

The *Forestry Act of 1959* came into operation on 1 August 1960, which was a time of mixed fortunes. Credit restrictions imposed by the Federal Government and the lifting of import controls virtually brought the industry to a halt. The second half of 1960 saw a record log cut of 141 million super feet, but the first half of 1961 saw this fall to less than half, the lowest since the war.

The conversion of rural leaseholds to freeholds was also having an effect. Although log production from private lands was falling rapidly, about 60 per cent of hardwood and cypress pine logs in recent years had come from private land. It was an indication that the land was not being developed in the way the government had intended. While the land was leasehold the department had some control over the standing timber, but it had no control once the land was freehold and it was clear that a great deal of timber was being removed.

Vic Grenning retired on 17 January 1964 after being head of the department for nearly thirty-two years. He had taken over a department that had been totally demoralised by the sacking of Swain and he had seen it through the war years and the period of reconstruction that followed. He had watched the area of plantations grow from barely a few hectares to more than 40 000, and the area of treated native forest had increased from 2000 hectares to nearly 300 000.

Grenning was succeeded by A. R. Trist, who had spent most of his life with the department. Trist was one of the department's first forestry cadets and one of the first graduates of the Australian Forestry School. Swain had then sent him to Yale and while in America he had studied southern pines in the belief that they might have relevance in Queensland. He also visited South Africa to study experiments in the use of these pines. Trist had been silviculturist and deputy conservator during most of Grenning's reign and so his appointment as conservator in 1964 caused little surprise.

Trist's silvicultural background was evident right from the start. In 1965 he wrote:

> Research work carried out by the Department has shown that great gains can accrue from the use of plantation stock from known parents of outstanding form and vigour and with good wood qualities. As a result of this work seed orchards of the major species being planted have either been established, or are in the process of establishment. It is with some pride that it is recorded that the entire plantings of some 3,000 acres of Slash Pine projected for planting in 1965–66 will be with stock produced from the slash pine seed orchard in 1963–64.

Two years later the Federal Government, which was now well aware of the future need for plantations, passed an enabling Act to subsidise increased planting of softwoods by the states. Under this Act, Queensland received an initial allocation of $201 000 and, as Trist wrote: 'For the first time in the history of forestry in Queensland, officers of the Department can feel optimistic about the prospects of achieving the objective of growing, within Queensland, the bulk of the State's requirements in forest products.' The result was a record planting of over 3000 hectares.

Curiously perhaps, this Act led to the first mention of conservation in the department's annual reports, fleeting though it was. It noted that the delays in extending the Act 'relates to concern over the impact of these planting programmes on the environment in each of the six States.'

This concern hardly seemed a direct threat to Queensland, and in any case the department, now headed by C. Haley, had embraced a policy that was, to some extent, a reflection of changing community attitudes. This was the policy of multiple use in which the public was encouraged to use forests for recreational purposes. The aim was partly to take some of the strain off National Parks and also to 'provide for some activities which are not permissible on National Parks'. Unfortunately these activities were not defined.

Multiple use was one of the buzz words from the 1970s on and it was embraced with fervour. The policy was not restricted to recreational use of the forests but extended to any other uses that could be made without interfering with forestry activities, such as beekeeping, quarrying and grazing. Many had been going on before, but there was now a policy to actively encourage them.

In 1975 the State Government passed legislation which divorced National Parks from the forestry department. This Act created the National Parks and Wildlife Service as a sub-department of the Department of Lands and the Service now took over the administration of all National Parks and the responsibility of creating new ones.

At the time of the change there were more than a million hectares of National Parks, about 500 kilometres of graded tracks and about two million visitors each year. It was a formidable achievement by the forestry department over the past seventy years and one that did much to support the view within the department that their record as conservationists was nearly perfect.

Following this separation the department undertook a massive internal reorganisation that was, as the next Conservator, W. Bryan, explained, largely a result of changing attitudes in the outside community:

CHANGING NEEDS, CHANGING VIEWS

Harvesting hoop pine thinnings at Nanango in 1973.

Greater public interest in environmental matters dictates that the Department increase and co-ordinate its activities in these areas. While it has followed for very many years a responsible policy of environmental concern with regard to its field of activities, it now faces a need to secure factual data to justify its management decisions in the face of often poorly informed and narrowly based criticism. In order to secure this data a separate unit has been established to provide and co-ordinate the necessary technical expertise in the many disciplines concerned. Environmental impact studies are being carried out with respect to significant departmental activities with initial emphasis on the major exotic planting programme. A range of environmental guidelines is also being prepared to assist staff in field operations, and guidelines were produced during the year for the retention of native vegetation within exotic pine planting areas to ensure continued provision of floral and faunal habitat for species indigenous to the locality.

There are two important points implied by this statement. One is that the changes came about because of public pressure, not because of any perceived need within the department. The other is that none of these activities had been practised until now. The last is not quite true. Individual foresters often took them into account, for foresters are certainly not insensitive to what is going on around them. But they had not been required to do so as a matter of policy.

Whatever the effect on the forest environment, the changes

associated with this policy certainly had a big effect on the department. There were seminars and workshops, mission statements, organisational development and all the paraphernalia of modern management. It was as if the department had suddenly found 'management' and had embraced it with the fervour of the newly converted.

Meanwhile, and almost unnoticed, for the first time the cut from plantations was greater than any other source. In 1979 a plantation of hoop pine that had been growing at Imbil for fifty-five years was clear felled and replanted to see if there were any unexpected problems with a second rotation. When it was felled the plantation produced nearly eight times the volume of the original, natural hoop pine stand per hectare.

If Bryan thought that his reorganisation and environmental guidelines would satisfy the conservationists he was wrong. By now conservationists throughout the country were very articulate in their concern for the forests and what they thought foresters were doing to them. Bryan was unable to understand why. Indeed his sense of outrage was clear in this report from 1980:

> In spite of the application of such environmentally sound management principles to all forest types, there remained a continuous conservationist campaign for restriction or even cessation of native forest logging, particularly in rainforest and associated wet sclerophyll hardwood forest type. Rainforest in Queensland is still the best preserved major forest ecosystem, with about 64 per cent reserved as State Forest and Timber Reserve and about 23 per cent as National Park. Importantly, approximately half of the rainforest area that has been reserved as State Forest and Timber Reserve will never be logged and will remain virgin for environmental and other reasons.

He was missing the point. Conservationists thought rainforests were too important to be at the whim of the Conservator. He might say now that half will remain virgin forests, but a future Conservator might decide otherwise.

For the forestry department, worse was yet to come. The conservationists mounted an all-out campaign to protect the north Queensland rainforests from logging and this came to a head on 5 June 1987 when the Prime Minister, Bob Hawke, announced that the federal government would proceed to nominate the wet tropics of north-east Queensland for World Heritage listing. It made the nomination in December and the following month the federal government introduced regulations banning logging under an

An aerial view of Kenilworth Forest Station in 1974.

external treaty and supporting legislation which it had passed in 1983. All this was done without involving the Queensland State Government.

The department mounted a strong campaign of defence. They pointed to the huge size of the area concerned. This was just under 9000 square kilometres, or the equivalent of a piece of country stretching from Caloundra to Coolangatta and west as far as Ipswich. They said that contrary to some conservationist statements, logging in rainforests did not involve clearing. Only about eight to twelve trees were removed from a hectare of forest and these were individually selected by forestry officers. One spokesman told the *Courier Mail*: 'Provided sufficient area is set aside for national parks, scientific, stream protection, recreation, scenic or other preservation purpose (which there is), there is no good reason to destroy an industry by banning timber production on the twenty per cent of the wet tropical rainforest that has already been selectively logged at least once, and some of it two or three times.'

This was the first major confrontation between the department and conservationists. Conservationists had already challenged the

State Government over its decision in 1983 to build a road from the Daintree to Cape Tribulation but that had not directly affected the department. The conservationists lost that battle, but they succeeded in obtaining World Heritage listing for the wet tropical forests of north-east Queensland and logging there came to an end.

For the first time the department had been seriously challenged over its use of the forests. Its initial reaction had been one of outrage that its professional competence should be questioned. This was followed by attempts to explain that forestry was a highly skilled profession which preserved forests for future use while continuing to supply the community with timber that it needed.

The department lost, and it was always going to lose.

Conservationists were generally middle class, educated and urban, although some had left the cities to become new settlers in the bush. Most had tertiary education and many were professional people, business people, creative people, and many were in the public service as teachers and academics. They were articulate, analytical, good debaters, and they knew how to handle politicians and the media.

Many people in the timber industry were not at all like that. Most workers in the forests had only secondary education, they had left school as soon as they could (often through economic need) and relied on their practical experience to stay in work. Many industry leaders came from a similar background—they were simply more capable and more enterprising. On the whole they were practical people who had no experience of public debate and who mistrusted educated theorists. Nor, incidentally, did they have a strong union to help them.

Those who did have an education worked for the forestry department and many had received specialised education as foresters. Some were articulate, some were not. Some were as skilled in public debate as the conservationists, others were not. But they were all restricted for two reasons.

One was that they were too inbred. They had much in common, they held the same things dear and there had never been much reason to question them. Their attitude to conservationists at first was to think of them as amateur if well intentioned meddlers, and then to complain that they spread untruths, which was partly correct.

The other restraint was that as public servants they could not act as individuals, no matter how competent some of them were. While leading conservationists appeared frequently in the media and became familiar figures, foresters rarely did so. They remained

Peeling pine logs for plywood at Austral Ply Company, Corinda, in 1974.

anonymous and unconvincing spokespersons of a government department.

As if this were not enough, the conservationists had all the appealing arguments. They wanted to preserve forests, while the industry, they said, wanted to butcher them for a profit. Even their images were better. Pictures of sunlit forests and appealing flora and fauna had an emotional attraction that was streets ahead of anything foresters could offer in defence. Conservationists could also draw on their own image: virtuous, concerned, caring yet free from the profit motive.

Conservationists used all these with great skill and with telling effect. Which is not to say that conservationists were always popular with the public. The excesses of some individuals, the possibility of violence and confrontations with police were unacceptable to many besides foresters, but they were relatively minor blemishes in what was otherwise an almost perfect campaign.

The main flaw in the conservationist argument was the economic consequences of locking up forests, and especially how those consequences would affect industry workers. It was obvious that people would be unemployed and entire communities might die, but this was largely ignored by conservationists. When they did confront this problem it was with two weak arguments. One was that the industry was in decline anyway. The other was that it was a political matter and politicians would have to find the answer. And so they did. In north-east Queensland the Federal Government offered a compensation package of $75 million to prevent direct and indirect unemployment.

Foresters and the industry relied on one main argument as a defence. It was that, contrary to the assertions of many conservationists, forests regrew after they had been logged. Indeed, many areas loved by conservationists, including the wet tropical rainforests, had been logged many times and their natural beauty was still impressive enough to warrant conservation. It was a good approach but it failed because it was a rational argument and that was not what these debates were about. In any case, said the conservationists, regrowth might bring back trees, but that did not mean that it brought back the other plants and fauna.

Nor did foresters and the industry state their case strongly or well. Instead they were rattled by the abuse that was thrown at them. They were not forest butchers and it seemed incredible that anybody could think they were. Foresters especially were stung by this accusation and pointed out that if it had not been for foresters there would have been precious few forests left for conservationists to fight over. The forestry department had been established to preserve forests when most of the community wanted to remove them for settlement. They were the first conservationists, they said, and they had been looking after the state's forests ever since.

This was true, but only up to a point. Certainly the department had been formed to preserve the forests, and it had done a good job ever since. However they had been conserving them for later use, not for aesthetic reasons. Indeed that is exactly what was expected of them, but it was not conservation in the current use of the word.

Conservationists introduced a completely different concept of what forests were for. Foresters accepted without question that they were there to produce wood, which was an essential resource. The conservationist view was that forests had other, more important uses. They constituted a unique ecology that was only now beginning to be understood, they were philosophically uplifting, they looked beautiful and they should be preserved.

It was a valid view provided the consequences were recognised, and provided that view was held across the entire community, not just a part of it. This view was so novel and unexpected to foresters that they had no answer. Their concept of productive forests had never been questioned, let alone challenged, by anybody within forestry.

The conservationist view questioned everything that foresters held dear. It said, in effect, that the visual and ecological qualities of certain forests were so high that they exceeded their commercial value. Leave them alone so that we and future generations can continue to enjoy them for what they are.

The battle over the tropical rainforests was fought again in 1990, this time over Fraser Island. The following year an inquiry headed by Tony Fitzgerald QC recommended an end to logging on the island even though it had been logged continuously for ninety years.

While all this was going on, the department continued with a seemingly endless succession of reorganisations. The Savage Report in 1986 encouraged government departments to develop a corporate management style, to simplify regulatory procedures and to develop leaner and more efficient management structures. The department hurried to obey and it examined nearly every aspect of its operations. Program management was introduced in 1988 so that the department was 'able to describe the program structure adopted and report for the first time along program management lines'.

The organisation was completely restructured. The three previous divisions were changed to four new divisions and branch managers reported to the head of each division. The six Director positions were scrapped and the Executive now consisted of the Conservator and the four heads of divisions. By this time the very mention of the word restructuring was enough to send staff running for cover.

It was, though, certainly time for a change. The department had become ultra conservative and rigid in its methods.

In the 1930s the department started to issue circulars to field staff about every aspect of field work. These circulars gave clear instructions about how any job was to be done and field staff were expected to follow these instructions to the letter. In many respects this was a sensible approach. Field staff were not highly trained, but there were many highly trained experts in Brisbane whose research and conclusions had to be incorporated into the field work as soon as the benefits were clear. The circulars were as good a method as any and certainly removed the need for field staff to make judgments. Whatever had to be done, there was a circular describing it.

But they also removed initiative. Even as the training of field staff improved, and indeed for many years, there was little field staff could do by way of input to the system. Their close observations of the results of these methods in their area were not accessed, so that regional variations went largely unnoticed for much longer than they should have.

Camping in the Booloumba Creek State Forest Park.

At the same time they gave the field staff a certain amount of security. Provided they had carried out their instructions, no matter how stupid they might seem to them, their jobs were not in danger.

The result was a formal, highly structured and rigid system of forest administration that had long outlived its usefulness. It was also reinforced by a personnel system that was not unlike that of the armed forces. Senior staff expected to be obeyed without question, and usually were. One district forester, which is the most senior position outside head office, was once told by his boss from Brisbane to get his hair cut.

The new approach was to move decision-making to the lowest level capable of making it. This varied with the decision to be made but the effect was that decision-making flowed down the management structure rather than being a preserve of head office.

These changes, necessary though they were, had an unsettling effect on the entire department. The changes were many and they all seemed to be happening at the same time. People were moved from one division to another, then all the divisions were reorganised, and on it went. The effect on morale was catastrophic.

Indeed this lack of stability was noticeable at the very top of the department. Bill Bryan left as Conservator in 1981 and was replaced by Jim Smart. In 1985 Smart was replaced by John Kelly, who in turn was replaced by Tom Ryan in 1988. It was a marked contrast to the period between 1918 and 1964 when the department had seen only two chiefs, Swain and Grenning.

Certainly by this time the job as head of the department had less and less to do with forestry. It had become an administrative position which called for sound planning along business lines and the ability to get along with the politicians of the day. Indeed the department is perhaps now at the stage where it does not need a forester as its head. There is an abundance of foresters at all levels within the department, but the administrative demands now placed on the head of the department call for different skills.

Following a decision in December 1989 by the newly elected Labor government the department became the Queensland Forest Service within an expanded Department of Primary Industries. In 1992 the *Primary Industries Corporation Act* removed the title of Conservator of Forests from the *Forestry Act* and passed all associated statutory responsibilities to the Director-General of Primary Industries. Tom Ryan, who left in 1993, was the last Conservator of Forests in Queensland.

So once again the department lost its independence and this time it lost the distinctive title of its head as well. It was all so traumatic that one began to wonder who was out there growing trees.

The aim of the department, however, had changed only slightly, and that was in response to recent events. It read: 'The overall purpose of the Queensland Forest Service is the sustainable production of forest products and services within a balanced conservation framework, which includes the multiple use management of State Forest lands, in accordance with the long-term interests of the community. In spite of the dramas of the last two decades, that aim sits comfortably with the foresters of today.

Mark Peacock
Forester

As a young forester, Mark Peacock is virtually at the start of his career. He is also one of the people who will take forestry well into the next century.

He was born in Sydney in 1968 and went to Scots College in Bathurst and then to Barrenjoey High School at Avalon Beach in Sydney. When he was sixteen the family moved to Nambour in Queensland. While he was there he learnt of the Forestry Training Centre at Gympie and set his sights on becoming a forestry trainee. He was attracted to the outdoor life and this was strengthened when he did work experience with the NSW Forestry Commission at Bathurst. 'I spent two weeks with a couple of foresters and we did a lot of things and went to a lot of places. I liked the variety.'

The family moved back to Sydney and Mark Peacock's Higher School Certificate was good enough to take him not just to Gympie but to the forestry school at the Australian National University (ANU) in Canberra in 1986.

He was there for four years and he loved every minute of it. 'It was without a doubt the best four years of my life.' He worked hard and enjoyed the life of the community. 'The whole forestry scene at ANU was a very close-knit group. You knew your fellow students and your lecturers extremely well and we got on really well.'

He handled the demanding course without much difficulty. The problems started when the course finished. The students in the years ahead of him had all been able to obtain jobs before they finished the course, but when it was his turn it suddenly became much more difficult. Only one or two students were lucky and the rest found themselves as newly graduated foresters with nowhere to go. Mark applied for a number of positions with the New South Wales Forestry Commission but with no success.

'Things were looking pretty desperate because it was getting to the time when university was about to go back and I still didn't have a job. I was about to re-enroll to do honours or a graduate diploma when a friend of mine in the Queensland department rang to tell me they were looking for a forester.'

He rang the department the same day, a Friday, and was asked if he could turn up the following Monday. 'I asked if that was for an interview. No, he said, that was to start. So I went from having nothing to being fully employed. Just like that. I was very lucky.'

When he arrived in Brisbane nobody seemed to know what to do with him. He did a few odd jobs for a while and was then sent to Dalby to carry out some inventory work, which he enjoyed. He then had a phone call telling him to pack his bags and go to Ingham.

He was to be there for two years, and if ANU was a great experience, doing forestry in North Queensland was even better.

'The work was extremely satisfying and very challenging. In the end I was the forester at Cardwell and responsible for a staff of about forty-two and a budget of about $3.5 million. We were doing a lot of projects and spending a lot of money. We were moving forward very quickly. It was really highly paced, challenging work.'

And very varied. He had to liaise with the Wet Tropic Management Authority, he was responsible for a number of picnic and camping areas, he monitored permit conditions and the operations of white-water rafting companies on the Tully River and he was involved in establishing new plantations. It was an active programme for someone who was barely twenty-three. 'I was very fortunate because the ranger and overseers were extremely capable people. I guess I was there just to give them direction and assistance when they needed it. They had most of their operations under control.'

He was also realising that he did not know as much about forestry as he had thought when he left ANU. The first twelve months were the worst: 'writing a memo was a whole new learning experience'. He also made mistakes, which was another part of the learning process.

Like many others he found the scenery magnificent and could not think of anywhere in Queensland he would rather work. 'I would occasionally sit on Mission Beach with fish and chips for lunch and think there weren't too many foresters in the country who would be having lunch at a beautiful place like this.' He was also impressed by the tremendous spirit amongst forestry staff. They had their own social clubs which laid on huge Christmas parties and organised social cricket matches.

It was with some regret that Mark Peacock left North Queensland in 1993, but it was another step in his career as he had been appointed Assistant District Forester in Brisbane. He is responsible for two forest nurseries, is involved in the planning of recreational facilities and administers the tree care programme. There is not as much activity as there was at Ingham nor does he spend as much time in the forests as he did.

He is conscious that the role of forestry is changing. The trend to commercialisation is introducing new requirements and he sees little point in trying to resist them. Changes are also being brought about by an increased awareness of forestry's responsibility to the community as a result of conservation.

He happily describes himself as a conservationist and finds no conflict in being a forester as well. In North Queensland he was involved with the preservation of the cassowary and occasionally objected to proposed freeholding of land on the ground that it would disturb the bird's habitat. He was also involved in the preservation of the rufous owl, and the plantation boundaries at Ingham were significantly modified for the benefit of this important bird.

'I think forestry does a lot that is beneficial to the environment and very little that is detrimental. I think in the balance we do a pretty good job as far as conservation goes but it is not widely recognised. The public perhaps doesn't hold that opinion, or perhaps they don't have an opinion at all and they are easily swayed against forestry. That is what we are up against. I don't think logging destroyed Fraser Island, it will be the tourists. They will be the real test.'

Mark Peacock is studying for his MBA by correspondence with the University of Central Queensland, and recently started learning Chinese at the local TAFE college because he did not want to go through life knowing only one language. He seems to be having little difficulty with either course.

Above all, he is committed to forestry. When most would see only a group of trees, this young and dedicated forester describes them in terms that would have impressed Swain.

This is dry sclerophyll forest—not highly productive. It has some reasonable trees and some reasonable ones have been taken out. Species are characteristic of a slightly hard site, that is, with relatively low fertility in the soil and with low rainfall. Species are grey gum, ironbark, red bloodwood and acacia. Grey gum and ironbark are dominant. The ironbarks are probably fifty to sixty years old. It looks as if it was prescribed burned about five years ago. Some ironbarks are suffering dieback. Hard to say why but could be insect attack or old age. This is typical open forest . . .

PART II

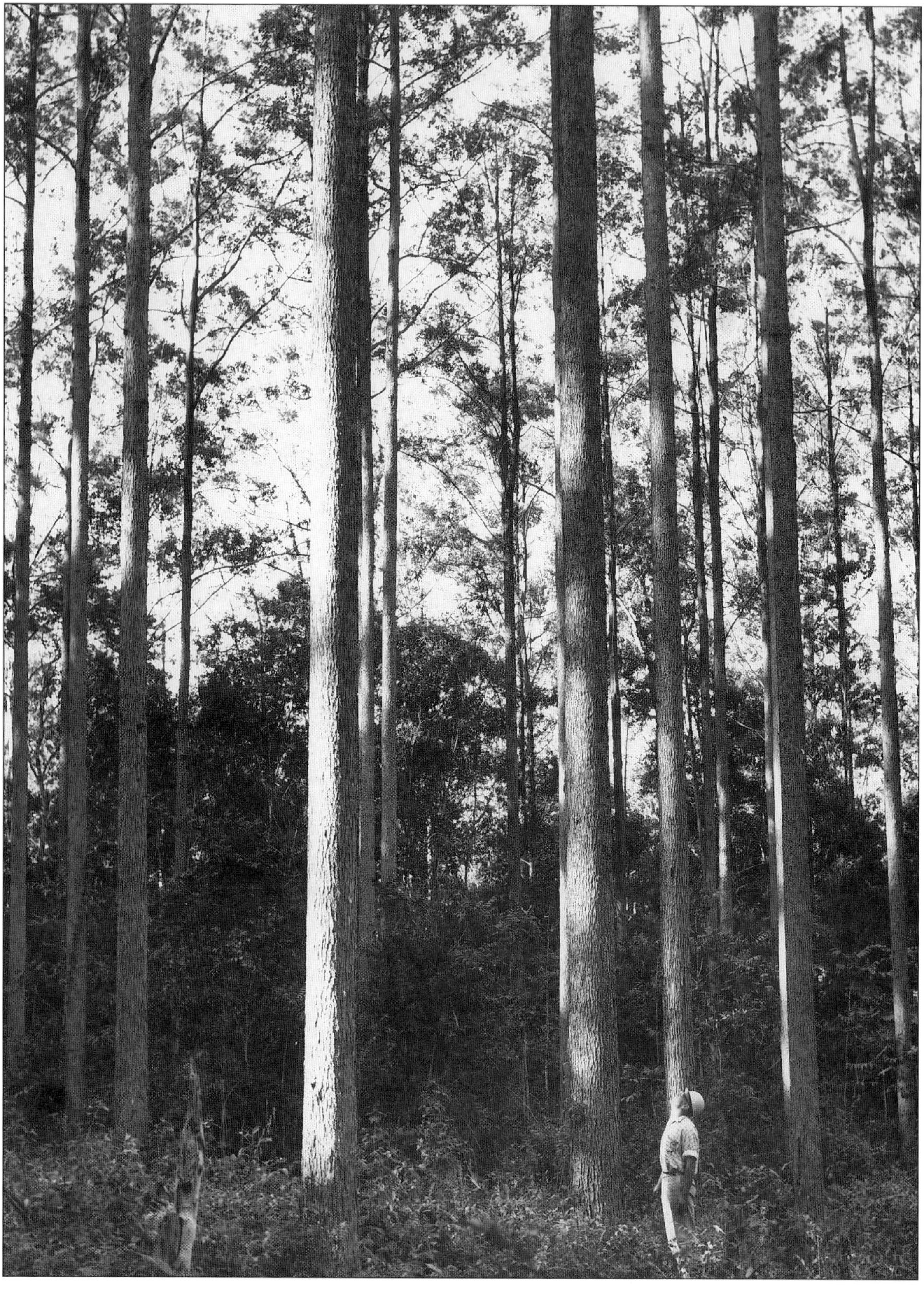

8 Managing Native Forests

The obvious question which must be asked is why manage native forests at all?

They are part of the environment and the fact that they are there at all is an indication that they are doing perfectly well without any interference. So why interfere when nature seems to have arrived at the best solution?

The reason is that what nature has produced might not be what people need. From time immemorial what people needed from forests was wood. But not just any wood. People needed wood for specific purposes, and these purposes were met by some trees better than others.

If the need was for fuel this was fairly easily met. Most trees will burn, and although some burn better than others the need for fuel could be met by a fairly broad selection of trees within a native forest.

But needs were often much more sophisticated than that. Wood might be required for building, for furniture making, for ship building, for constructing wharves or building bridges. Each of these more sophisticated uses could be met by only a limited range of trees.

Wood for ship building or for use in wharves, for example, has to be able to resist attack by marine borers, which not all wood can do. Wood that has this ability comes only from certain species of trees. Similarly, wood used in building has to be strong, needs good

Opposite: *An example of a well-managed native forest in south-east Queensland.*

mechanical properties and must not decay when exposed to the elements. Wood that combines these properties comes only from certain species.

So while native forests might be in perfect tune with nature they might not be in tune with the community's needs. They might produce abundant wood but it might not meet a need, and even if it does availability will be haphazard.

One of the skills of forestry, and it is an ancient one, is to manage the forest so that it produces an abundance of those species of trees, compatible with the environment, that meet specific needs. Thus England in the Middle Ages had an almost limitless need for oak for building and repairing ships. Foresters therefore managed the forests so that they produced large quantities of good oak to meet this need.

Managing forests in this way is far from easy and it is not surprising if early techniques seem crude now. There are two basic problems. The first is the environment, and specifically the nature of the soil and the climate. Most species of trees will thrive only if their specific needs are met by their environment, and in some species these needs are very particular. In order to manage, or encourage, the forest to produce these species the forester had to know what their needs were and then try to meet them.

In doing so the forester ran into the second problem. Because trees take a long time to reach maturity the forester had to wait a long time to discover whether he had been successful or not. Indeed, because it might take thirty, forty, fifty years or more, the forester who began the project might never know the outcome as his life span was shorter than that of the trees he was trying to encourage.

Improving livestock such as sheep or cattle is, by contrast, much quicker. If selection of breeding stock is well based to support the required improvements then the results will be seen after only a few generations, each taking no more than a few years. Each generation will also provide an increase in the number of suitable breeders, so the development base always increases. The result of this has often been spectacular. In not much more than a hundred years Australian sheep breeders produced sheep capable of carrying a fleece that would have crippled the sheep that Macarthur knew. A development of similar magnitude in forestry would have taken many centuries even if it were feasible.

The forestry skill of managing forests or, more specifically, growing trees is called silviculture. It is the corner stone of forestry and, until the recent development of plantations, silviculture of native forests had been the essence of forestry from the earliest times.

Disturbance to rainforest by natural events can be quite extensive as the cleared area around this collapsed satin sycamore shows. However, the forest usually recovers rapidly.

Swain, who did much to advance silviculture in Queensland, described it thus:

> Sylvicultural practice is founded upon intimate knowledge of the life-histories of trees. The life history of every species is different, and must be observed, investigated, and recorded patiently for each, from the seed to maturity and decay. Whilst Red and White Cedar and Maple spring exuberantly from the ground only to meet in frost, drought and twig borers, the very dangers to which their undue succulence exposes them, the minute Hoop Pine seedling lurks unseen beneath the weed masses, and by its ability to endure in babyhood an almost overwhelming shade, and later to resist the blight of frost and drought which lays low its fast-growing competitors, succeeds to ultimate dominance of the forest . . . All these processes and a myriad more must be isolated before sylvicultural command over species may be established, and in no State of

Australia are the difficulties greater or the sylvics more complex than in the sub-tropical State of Queensland.

In other words, foresters need to understand the growing requirements of the species they wish to encourage and then, within the limits of nature, meet them. It can be easily stated, but it is not easy to do.

In 1876 a small experimental trial was started on Fraser Island to study the artificial restocking of depleted forests as well as the processes of natural regeneration. Six years later the trial was well established and it was the start of silviculture in Queensland.

To help nature restock forests that had been cut a gang of workers cleared competing undergrowth from over 4000 kauri pine seedlings that were growing naturally. In addition, more than 16 000 kauri were transplanted from their original locations into existing openings in the rainforest as well as in strips cut into the forest. Others were transplanted into newly established nurseries for future use and seeds of a number of species from other Australian states were also planted.

By 1898 this experiment had failed. At the time it was thought that this was because the sandy soil was not rich enough, but it was probably due to the natural vigour of the forest and the competition from weeds.

In North Queensland cedar had been found regenerating naturally in gaps in the rainforest and in 1902 experiments started at the Kamerunga nursery in Cairns to test cedar cuttings as a source of planting stock and the possibility of transplanting seedlings from the road sides to openings in the forests. In the next two years cedar seed from Imbil was also raised in the nursery along with a few other species.

In 1913, eleven years after the start of the experiment, the effect of the twig borer was first noticed. An earlier attack on artificially grown cedar on private land near Maryborough had been recorded but it had not been identified at the time and had been long forgotten.

It now became clear that the twig borer was devastating to artificially raised cedar. By 1916 it had ruined most of the plantings in North Queensland and yet another early experiment came to an end. There was, however, some success at regenerating maple. The forest was cleared of non-commercial timber and the floor was also cleared. This usually resulted in good strikes of maple seedlings.

Meanwhile there had been further experiments on Fraser Island. In 1913 'improvement fellings' were carried out on 200 hectares of densely stocked hoop pine. The purpose was to thin out the stand in order to provide enough space for the remaining trees to develop.

Another technique introduced at this time was 'regeneration felling'. This meant clearing an area of the forest and burning the forest floor to produce a very fertile seed bed. Seed from other species was then sown (and this was particularly successful with cypress pine) or, in the case of blackbutt, seed fell naturally from trees that had been retained for the purpose.

In 1913 improvement fellings took place in the Brisbane district, first on hoop pine scrubs and then on hardwoods. Although still called improvement felling the technique when used with hardwoods was quite different. These forests were logged of marketable timber and the remaining trees were ringbarked and the debris burnt. The forest was then left alone so that it could regenerate naturally. Until 1920, then, silviculture in Queensland was largely experimental and on a small scale. There was little labour available for this kind of work and funds were scarce. Although there was little sophistication the thinking behind these early attempts was well based.

It was Swain who brought silviculture in Queensland to a refined skill. His knowledge was already beyond the experimental stage and he quickly established silviculture as an essential and productive technique in Queensland's forests. By 1923 he had established a single technique which was used to increase wood production from the forests, although the technique was modified slightly to meet local conditions. This technique consisted of three distinct stages.

In the first stage the veterans of undesirable, that is non-commercial, species were ringbarked. In the second stage the forest was heavily logged of marketable trees with no restrictions on girth size. With the forest now much more open, the third stage involved clearing the remaining undergrowth with axes and hooks. The slashed growth, together with the fallen debris from the earlier ringbarking and the butts and tops from the logging, were burnt the following summer.

This burn had two effects. It produced a clean seed bed rich in ash, and it encouraged the trees that had been left standing to empty their seed capsules, now dry, onto the prepared ground. If the burn was successful, some remaining seed trees were ringbarked and their seed capsules were released to the ground as the tree died. A fourth stage was sometimes carried out later. In this the young seedlings were thinned out and any gaps were planted with eucalypts or other species.

Most forests could be treated in this way but there were some variations. Hoop, maple and even yellowwood were usually logged and this was followed by liberation treatment of the younger denser stands. Cypress was heavily logged and the remaining stems thinned so that they were about 1.5 metres apart. There was no burning with these species because they had little or no resistance to fire.

Under Swain's leadership this technique was applied in the 1920s to forests at Fraser Island, Dalby, Atherton, Benarkin and Goodnight Scrub as well as in coastal forests between Brisbane and Gympie, usually with noticeable success. The main problem was the loss of seedlings during summer droughts but the success rate was generally very high.

Much of this success depended on the judgment of individual foresters. They had to decide which trees to ringbark, which to log and which to retain, and then make sure that when the debris was burnt the fire did not damage the remaining stems. It is not surprising, then, that Swain saw the need for trained staff. This technique could be successful only if people knew what they were doing.

In the 1930s the area of native forests treated in this way increased enormously. By 1935 more than 57 000 hectares of mostly eucalypt and cypress forest had been treated and two years later this had risen to over 91 000 hectares.

By the 1930s the technique had been simplified. Most hardwood and cypress forests were simply logged, the useless stems destroyed and the remaining good stems thinned. Regeneration burning now seems to have been restricted to Fraser Island, presumably because of the blackbutt there.

There was another development at this time which was also to have a profound effect on the management of forests. This was the introduction in 1937–38 of 'economic tree marking'. Until then a sawmiller had been granted rights to a specific area of forest and a forest ranger then marked the trees to be felled. The difficulty arose because the ranger included trees that were to be removed for management purposes as well as those that were merchantable. The cutter, however, was interested only in trees that had a commercial value and so he felled those and ignored the rest that had been marked.

It was common for the ranger to call the cutter back into the forest and insist that the remaining marked trees be removed, even though the cutter protested that the work would show no profit. Economic tree marking recognised this for the first time. Under this new system the cutter removed all the marked trees but he was guaranteed a minimum payment for trees that proved to be defective. This payment was met by the sawmiller and the department.

This simple and fair solution had three advantages. It produced the best use of the stand, it meant that silvicultural requirements

A forest ranger marks a maple silkwood for felling in rainforest on the Mount Windsor Tableland. The benchmarks show the direction in which the tree should fall. The rainforests of Far North Queensland are no longer being logged.

could be blended with commercial needs, and it meant that follow-up treatment in the stand would be effective. The system was in use for over thirty years and led to a considerable improvement in the quality of the stands.

Meanwhile, by the end of the 1930s large scale silvicultural operations were being carried out in at least twelve districts from Brisbane north to Bundaberg and west to Dalby. The time had also come to treat for the second time those forests that had originally been treated in the early 1920s.

By this time there was a considerable body of silvicultural knowledge based on research and experience and this continued to expand. The technique was modified in the light of this knowledge but the basic concept of encouraging natural regeneration is still valid today and the techniques are recognisably similar.

This technique, however, was not generally applied to rainforests because of the nature of the forests themselves. Typically, they consist of a great many species of trees. A few were highly sought after as cabinet woods, a larger number were used for more general purposes, and an even larger number lacked any commercial qualities, or so it was thought.

Logging unmerchantable trees and then favouring desired species was clearly impracticable in a rainforest. There would have been so little left that the nature of the forest would have been totally changed if not actually destroyed. Thus the first steps in the management of rainforests were very modest and consisted of enrichment planting in which desired species were introduced to areas of the forests which were deficient in them.

Attempts were also made to encourage the number of species being used. This did not make the forests more productive as the species were already there, but it was an attempt to avoid excessive demand on the few species that were used. In the early 1900s only about ten rainforest species were being used and numbers remained small until the Second World War, when the general shortage of timber led to the use of more than a hundred rainforest species. By the time logging in rainforests came to an end over 150 species were in use and this exceeded by a large amount the number of rainforest species in use in most other parts of the world.

Attempts at enrichment planting in rainforests were overtaken by the need to establish plantations. Although attempts were made to establish plantings of rainforest hardwoods during the First World War these were generally unsuccessful and in the 1930s plantings concentrated on establishing the native hoop pine on cleared rainforest sites. By the start of the Second World War a little over 500 hectares of these plantations, mostly of hoop pine, had been established.

This coincided with the increase in the number of rainforest species being used and it became obvious that clearing rainforests for plantations destroyed large amounts of advanced species which were now becoming merchantable. At the same time high mortalities in the hoop pine plantations caused by root rot fungus indicated that the plantations were unlikely to be successful and so the clearing of rainforest for this purpose came to an end.

However, the problems associated with selective logging of rainforest species remained. The main one was that selective logging did not provide the conditions necessary for their regeneration and as artificial regeneration had proved to be difficult, if not impossible, these desirable species were in danger of being lost forever.

The dilemma might seem curious now. The forest would

regenerate naturally only if a wide range of species were removed, thus retaining the balance that had been arrived at by nature. But while the market was more than willing to pay well for cabinet species it was reluctant to accept less desirable ones that could not compete in distant markets with timbers from local forests, especially as they might be available only in small quantities.

In an attempt to overcome this impasse, in 1946 the department introduced compulsory logging of about a hundred species and at the same time introduced a programme of silvicultural treatment in the rainforests. The object of this treatment was 'to induce a good representation of and/or to foster the growth of the better species by reducing the competition from weeds species and to do this as cheaply as possible'.

Obviously this assumed that the forest to be treated contained a large proportion of commercially useful species, otherwise eliminating the weed species would have amounted to the destruction of the forest. This programme was also supplemented by underplanting of desirable species where natural regeneration was inadequate.

Harvesting of the forests was controlled by tree marking rules introduced in 1955. These rules prescribed a diameter for each species measured 1.3 metres above the ground or above the buttresses and trees with greater diameters were marked for cutting and another mark indicated the direction in which it must fall to minimise damage to other trees. This method resulted in considerably less damage to the remaining forest than was usual in most rainforests in other parts of the world. The damage that was unavoidable posed little threat to the forest, however, as the problem was usually one of overstocking rather than the opposite.

Care was also taken to make sure that the canopy of the forest was not opened too much otherwise the introduction of sunlight would materially change the nature of the forest and it would be many years before the original balance was restored. Thus the canopy was never reduced by more than 20 per cent and usually by much less.

The tree marking rules also provided for the removal of damaged trees and those infected by fungus even though they were below the diameters indicated in the rules, but they did not allow the thinning of dense stands even when they were only slightly below the approved measurement.

In addition, other trees were marked for preservation instead of cutting. These were selected trees of prime species which were to be kept as seed trees. Trees in the vicinity were carefully selected so that these seed trees would not be damaged when they were felled, and if this was not possible then no trees were marked for logging in

Felling a large hoop pine at Yurol in 1940.

their vicinity. This treatment of northern rainforests was carried out with great vigour for the next decade or so and many hectares of forest were treated in this way. In the end, though, it fell into disuse. One reason was that because it was very labour intensive it was also very expensive. Another, more compelling, reason was that it did not produce the results that were expected.

Following the treatment there was usually some immediate improvement in the natural regeneration of desired species and in the form and vigour of young trees. But in the long term the benefits were difficult to detect. The forest had, it seems, its own resilience and in the end it exerted its superiority over human attempts to control it. By the 1960s the records showed little difference in productivity between those forests which had been treated and those which had not and so the programme was abandoned in the early 1970s.

It follows, for the same reason, that neither did it do any harm,

The policy of multiple use encourages recreational and other uses of State Forests.

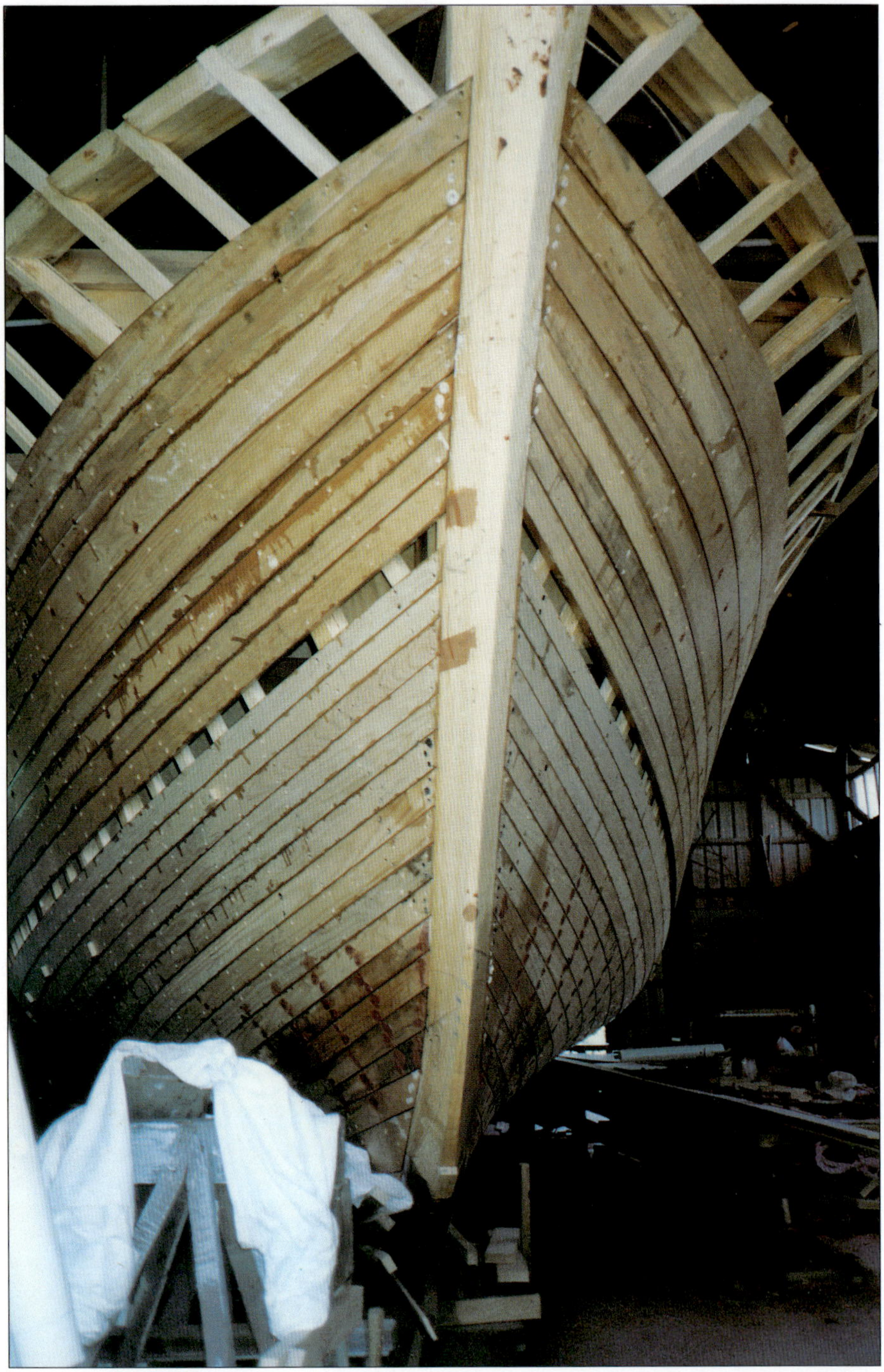

Above: *Rainforest at Woongoolba Creek, Fraser Island.*

Opposite: *A 46-foot motor cruiser under construction on the Gold Coast in 1994. The vessel is hollow-heel carvel built using hoop pine throughout. Photograph courtesy of Gary Doornbos.*

Grey kangaroos in the State Forest at Beerwah.

A large kauri log from Mount Windsor in Far North Queensland.

although by that time there was no shortage of critics. Even trained foresters were often unable to tell whether a patch of forest had been treated in the past or not, and when they did know, perhaps because they had been involved in it, they often listened to conservationists enthusiastically declaiming some twenty years later that that was how rainforest should be—pristine and untouched.

Another type of forest in Queensland that was singled out for specialised silvicultural treatment was cypress pine. Cypress pine grows in south-east Queensland west of the Dividing Range and its value was recognised from the earliest days. It often grows in forests containing eucalypts and acacias which generally do not provide logs of milling quality. Unlike those, however, cypress is resistant to termite attack and it soon became a popular building timber.

These forests were also used for grazing and in 1957 the government allowed certain grazing leaseholders to convert their rights to freehold. This effectively removed the cypress pines growing on the land from government control, although valuations made by officers of the department were reflected in the value of the transfer, and they could also argue against alienation in the Land Court if the stand warranted special attention.

Cypress pine regenerates well, indeed almost too well, and dense stands of young cypress pines are quite common. The trees in these stands are overcrowded and so they grow slowly and usually develop spindly stems and poor crowns.

The treatment of these forests usually follows commercial logging, which removes many of the mature trees. The dense stands of young trees are then thinned so that there are no more than about 300 per hectare. At the same time any non-commercial species that are overtopping or crowding these select stems are also removed.

This relatively simple procedure allows the select stems of cypress pine to grow into prime merchantable trees and at a faster rate than they would if they had to compete with their close neighbours. It is not uncommon for the growth rate to increase by up to ten times as a result of this treatment.

While silviculture undoubtedly increases the productivity of useful species in native forests there remains another fundamental in the management of these forests. That is the need to make sure they are not destroyed by overcutting. This is done by a practice called sustained yield and it was originally developed in Europe, where the need to avoid destroying forests had long been obvious.

Briefly stated, the practice of sustained yield maintains that the amount of timber cut in a forest in, say, a year is no greater than the total growth of the forest during that year. It is not that year's growth that will be cut, of course. The cut will come from mature growth but it will not exceed the amount of new growth during the same period. If this policy is followed the forest will never diminish and it will, in theory, continue to produce timber for cutting year after year.

In practice it is not as simple as that. Calculations of forest growth, let alone the prediction of future growth, are not easy and yet they are essential if this policy is to achieve its aim of preserving the forests in perpetuity. And even when these calculations and predictions are correct they can become unreliable through natural events such as storms, fires and insect attacks.

Another problem for managers of old-growth forests is that initially there will probably be no increase in growth in the forest because growth and mortality have arrived at an equilibrium. With no net increase the policy of sustained yield would imply no cutting. In this case, mature trees have to be removed from the forest so that regeneration can start. It is only when young healthy specimens develop that the forest begins to show an increase in growth rate which can eventually be harvested. It is not surprising that this procedure invites criticism because it does indeed seem to be attacking and destroying a forest. But the result is considerably increased growth in the future which can then be managed on the basis of sustained yield.

Another important aspect of managing native forests is selective logging. This was touched on earlier in the context of rainforests, but it is applied to all state native forests in Queensland. It means that trees to be removed from the forest are selected by a forester, and not the sawmiller who has the logging rights to that forest. The miller, or more usually the contractor working on his behalf, can remove only those trees marked by the forester, and they must remove all of them whether they are commercial or not.

As we have seen earlier, selection by the forester is very specific and combines the two needs of utilisation and silviculture. The suggestion that native forests in Queensland are clear felled for logging is quite wrong. There is virtually no clear felling of native forests in Queensland and, apart from land clearing for other use or in private forests, there never has been. On the contrary, every tree that is felled has been marked by a forester on the basis of sustained yield, the requirements of silviculture and the well being of the forest.

In the last few decades another element has been introduced to the management of native forests and that is the need to protect water catchment areas and the associated need to avoid silting of water courses because of erosion. Trees along water courses are now preserved in their natural state and other areas that are logged are done in such a way as to minimise the problem of erosion.

Associated with this is the policy of multiple use, so that forests are also used for grazing and beekeeping, for example, as well as providing areas for recreation. Some areas of native forests are not logged in order to maintain these alternative functions of the forests.

It can be seen, then, that managing native forests, far from being unnecessary, is a highly skilled activity and a very complex one. It might not always be successful, or to everybody's taste, but the result is a continuing supply of high quality wood to a community that still regards it as an essential resource.

Sam Dansie
Retired Forester

Sam Dansie is one of those remarkable people that once you have met them you never forget the experience.

Not that he would agree with that. On the contrary, Sam does not think he is at all remarkable. He is a quiet and rather private person who keeps his views to himself and says no more than he has to. But he is much more than that. He is one of the best bushmen in far North Queensland and his practical knowledge of the tropical rainforests is second to none. Nor is it academic knowledge, for Sam Dansie learnt it by spending the best part of a lifetime living and working in the scrub.

He was born on the Atherton Tableland in 1927. He spent three years in the forces at the end of the war and on his discharge he returned to the Tableland and bought a dairy farm. He ran this single-handed and even though he was fit and about twenty-one it nearly killed him. He sold the farm and in 1951 joined the forestry department at Atherton.

His first job was as a labourer treating rainforest. The war had proved that many lesser rainforest species were usable and in 1948 a policy of rainforest treatment was introduced to encourage them. The treatment consisted of selecting useful species and then liberating them from competing growth. This was done by brushing the undergrowth and ringbarking taller trees with an axe.

It was hard physical work done in very harsh conditions. A good labourer might do half a hectare a day, a bad labourer would not stay. Living was rough too. Mostly they looked after themselves in a bush camp.

Although the location might change, the work was pretty much the same. He went on to brushing rubbish out of a hoop pine plantation and then it was on to Baldy Mountain for more treatment.

In 1953 he joined the rainforest research section as an overseer, putting in plots and carrying out measuring. In 1959 he became ranger in charge of rainforest treatment, tree-marking for logging, and plantations. In 1966 he became senior ranger, harvesting and marketing and by the time he retired in 1988 he was marketing inspector—the highest position that could be attained by a person without academic qualifications.

Others have spent their working lives in the rainforests but few saw them with the keenness that Sam did. Nothing escaped him, everything had significance and he forgot nothing.

It was when he was working in marketing that he began to suspect that treatment was having little effect on the rainforest. 'It doesn't matter what you do with a rainforest, it will never produce more than its natural capability, which is between forty and sixty cube per hectare.'

He became convinced that rainforests knew how to look after themselves. He sees the attempts by humans to 'improve' them as being largely futile and presumptuous. In the end the forest does it its own way.

Although others also had doubts about treatment, the policy was a favourite of Trist, who was then conservator and who considered himself a rainforest expert. Nobody was willing to express these doubts to Trist, however, and the policy was not abandoned until Trist retired, by which time it was hopelessly uneconomic anyway.

Conservationists often portrayed foresters as 'butchers', ruthless people who exploited the forest and who cared little about the environment. It is a pity that these people do not meet Sam Dansie. Not only does he care a great deal about the environment, he knows it as only an expert can.

His first brush with conservationists came during their protest on the Windsor Tableland and he used it to try to bring home to the timber industry the need for them to accept environmental discipline and to treat the forests with respect. He did not have much success but his warning was certainly timely.

He usually got on well with conservationists and had much in common with them. When groups arrived in Atherton he took most of them into the forest while their spokespeople had meetings with senior forestry staff at logging sites. He would sometimes have twenty conservationists with him in the forest and he thought most of them were OK. Not many shared his opinion at that time in far North Queensland.

To the surprise of many people, Sam Dansie was largely in favour of World Heritage listing. Because overcutting had sometimes been allowed in the past in order to encourage regeneration the forests were not as well stocked as they should have been. Sam thought there were only two or three years of virgin cut left and if the area was listed then at least the industry would receive compensation. Without it there would eventually be no cut and no industry.

If you walk through a rainforest with Sam Dansie you soon realise that you are seeing it with an expert. He seems to know every tree, not just as a species but as an individual he has watched grow over several decades. He knows when the forest was last treated. He knows with certainty, because he helped to do it. Every plant, every insect, everything in the forest, he knows them all. This is his country.

Sam Dansie tries to express some of his thoughts in poetry, and always has. It seems the only outlet, for much of his thinking is deep and does not lend itself to conversation. This is one of the poems written by this very remarkable man.

On Mount Windsor when the day is done

The chowchillas sing their even song
While they run, the ground along
The dingo howls to attract its mate
While I sit by the fire, and contemplate
How the next day's goal be won
On Mt Windsor when the day is done

The tall forest grass
From the wet season past
Parts and waves as the wallabies pass
With the thump thump of paw and tail
As they head for the waterhole in the vale
And the roosting parrots shriek as one
On Mt Windsor when the day is done

The kookaburra laughs its final laugh
To bid the day farewell
While the breeder cow lows to its calf
Within the nearby dell
There's a tinkling of a distant bell
On some station bull, rebel

SAM DANSIE

As shadows grow with the setting sun
On Mt Windsor when the day is done

The kauri pines stand erect with pride
Growing upon the mountainside
Cast shadows broad as though to hide
Other trees that grow 'neath the spreading crowns
Brightly green with leaf and cone
On Mt Windsor when the day is done

The death adder stirs from its sleepy coil
To warm its belly on the dusty track
While the bandicoot digs in damper soil
With hope of finding a tasty snack
The green tree frog jumps as if it's fun
On Mt Windsor when the day is done

When finally the sun has set
And all the shadows completely met
To enshroud the mountains
In a complete dark net
There's the fluttering past of the little bat
And the shrieking snarl of the native cat
As they venture out 'neath the darkened hood
To search and find their vital food
And the crickets chirping one by one
On Mt Windsor when the day is done

9 Plantations

If the forestry department had not had the considerable foresight to see the need to establish plantations, then Queensland today would be importing most of its timber.

This need was recognised in the first few decades of this century even though the shortage of timber would not occur until some thirty years later. For those aware of the situation the problem was not immediate. Indeed it was unlikely to occur within their lifetime. But if a solution was to be found, then work had to start immediately. The solution was plantations, and they had to be planted long before the problem became universally visible.

While good silvicultural techniques dramatically improve the usefulness of native forests, there are limitations. The soil and climate will largely determine which species are present and while the performance of these species can be improved there is little scope for changing them. If these species are of little use as timber, the value of the forest in terms of production might be very little. Nor does the forester have any control over the terrain. The native forest might be remote and mountainous, in which case it might be impractical and uneconomic to use it as a source of timber.

But the demand for timber is not restrained by such things. Timber is needed in ever-increasing amounts as a resource and it is up to the forester to supply it. Explaining the problems of terrain and the limiting concept of sustained yield does not supply a hungry market.

PLANTATIONS

A soil survey crew boring to provide data on soil profiles prior to the establishment of a plantation near Bribie in 1958.

Because the limitations of natural forests meant that they would be unable to meet increasing needs for certain types of timber, especially those used in building, an alternative had to be found if supplies were to be maintained. That alternative was the plantation.

The concept was simple. An area of land was cleared and then a species of tree was planted that would thrive on that soil and in that climate. Then the trees were grown as an agricultural crop, much like any other. In due course they were harvested and the process began again.

The advantage was obvious. Market needs could be met because trees were grown specifically for that purpose, and grown to maximise productivity, in a more or less controlled process. If the native forests could not supply the market, plantations could, and would continue to do so by repetition.

The problem was the length of time between planting and harvesting. This was thirty, forty or fifty years, depending on the species, and during that time there would be little financial return. Private landowners were never likely to be enthusiastic over a deal like that, even though they were supplying the market with timber from native forests on their land.

The initiative could come only from the forestry department, but convincing others of this need was far from easy. One reason was that most people saw huge areas of timbered land around them and the concept of a shortage nearly half a century later was hardly of great concern even if it were true. This was particularly the case with politicians. Most thought no further than the end of next week. If foresters, by training, were able to think that far ahead, politicians, by their nature, were not.

As if this were not enough, there was another problem of equal magnitude: even within forestry there was no certainty as to how to establish and manage plantations. For centuries forestry had been about the management of native forests and much was known about that even though European and American knowledge had to be reconstructed in Australia.

But there was no store of knowledge about growing trees as an agricultural crop in plantations. This could be obtained only by trial and error, and that is why early attempts at plantation growing were hesitant and on a small scale.

Attempts were made on Fraser Island in the 1880s to grow nursery stock of various species to be planted in gaps in the native forests, but the results were disappointing. Further attempts were made in North Queensland in the early years of the century to raise nursery stocks of trees for enrichment planting in parts of rainforests that had been heavily logged. Although these early attempts were not successful they did provide a basic knowledge of nursery management, pest problems and the adaptability of species to different sites. This knowledge was to be very valuable in the early days of plantations.

In 1910 the Under Secretary of the Department of Public Lands, the parent body of the forestry department, wrote: 'It seems not improbable that in the not distant future the needs of the inhabitants of Queensland, so far as regards pine timber, will have to be met by exotic varieties of inferior quality, secured by importation in a manufactured state or from local plantations on land not capable of producing the indigenous varieties.'

This recognised that future needs for softwoods would not be met by the native forests and that pine plantations were a possible solution. The suggestion that the plantations would be of exotic species was in line with thinking at that time. Plantations would have to be established in areas that were not currently growing native pines, and therefore those areas would be unlikely to support them in plantations. They might, though, be able to support exotic pines, perhaps from America.

A small experimental plantation was established on Fraser Island in 1911 and in 1913 an experimental station was set up in Brooloo in the Mary Valley. The work on Fraser Island compared the

PLANTATIONS

Raking slash pine cones to release their seeds, Beerwah, 1968.

performance of native hoop, bunya and cypress pines with exotic pine species and some eucalypts in both assisted natural regeneration and their suitability for use in plantations. It was in the Mary Valley, however, that the merits of native pines for this use were first recognised.

In 1916 a forest station was established at Imbil and this included a nursery and arboretum. The following year two small areas were cleared at Brooloo and planted with hoop pine and other species to compare natural and artificial regeneration. Jolly, then director of the department, reported: 'Of the coniferous species tried, hoop pine proved by far the most hardy and satisfactory, its superiority over exotic species being very marked and its rate of growth for the first year being distinctly good, annual shoots up to 3 feet having been recorded.'

This encouraging performance of hoop pine was confirmed by similar results from hoop pine planted at Atherton in 1916. The only drawback so far was that these plantings produced a luxuriant growth of weeds as well.

In 1919 two hectares were cleared near Imbil and planted with hoop, bunya and cedar as well as a number of exotics and other species. After the trees were planted pumpkin seeds were sown in the hope that their prolific growth would suppress the weeds but they were destroyed by beetles.

When Swain took over the department he continued this experimental work and, as he did with silviculture, refined and elaborated it. By this time other states had seen good performance with exotic pines, but Swain had no doubt that in Queensland the native species were far superior to them: 'While the outstanding fact of Australian silviculture in the past has been the phenomenal planting success in the southern states of the introduced Californian *Pinus insignis* [radiata], the achievement of the Queensland experiments is a comparative triumph of the Australian species and the virtual rout of the exotics . . . the hoop pine of Queensland . . . has thrived under conditions which would kill *Pinus insignis* to a plant.'

The fact that hoop pine was so successful was of enormous importance because it was this species that was in such demand and which therefore was under most threat in the native forests.

Hoop pine grows in a wide variety of soils provided the annual rainfall exceeds 750 millimetres. It occurs naturally as scattered trees towering above the upper level of the rainforest and only rarely occurs in dense stands. Because of competition from other rainforest species most hoop seedlings fail to reach maturity.

Wood from hoop pine is exceptional. It is fairly soft, it saws and dresses easily and it takes stain well. Because of this it was in great demand for internal work in buildings such as flooring, mouldings, furniture and other forms of joinery.

Its name comes from the hoops that it leaves on the forest floor. These are bands of bark which form naturally on the trunk and can easily be seen on the growing tree. When the tree falls, the softwood of the trunk decays quite quickly and the tougher bands of bark are left on the forest floor.

In 1920–21 Swain, encouraged by several years of experimental results, decided to establish the first commercial plantations. During that season over 60 000 nursery-raised seedlings were planted on 48 hectares in three different regions: 26 hectares in the Mary Valley, 12 at Atherton and the rest on Fraser Island. Hoop and bunya pine made up 80 per cent of the plantings and the rest were scrubwoods, conifers and eucalypts. Nurseries were in production at each of these areas and a new softwood nursery was also established at Benarkin during the same season.

This work was made considerably easier the following year by the invention of the Weatherhead tube, which has already been described. This made it easier to raise seedlings in the nurseries and easier to plant them into the ground when they were ready.

In spite of the progress that had been made, however, the future problem had not yet been solved. In 1926 it was clear that the diminishing supplies of hoop and bunya from native forests would, at the present rate of cut, be finished in twelve years. The new plantations, on the other hand, were not expected to supply mature

PLANTATIONS

Loading an aircraft for aerial fertilising over Beerburrum.

timber until they were fifty or sixty years old. The solution, imperfect though it was, was to reduce the present rate of cut from native forests and to rapidly extend the area of plantations, perhaps using faster growing exotics.

By now there were about 500 hectares of plantations and Swain set an annual target of 2000 hectares, which meant that practically the entire resources of his department would be directed towards plantations.

The next question was where these new plantations should be established, especially as there was a general attitude at that time that forests should be removed to make way for settlement. Fortunately the department had managed to obtain nearly 2500 hectares of wallum land near Beerwah, land that was so poor that even the most enthusiastic settler looked elsewhere.

An experimental station had already been established at Beerwah in 1924 and Swain decided to use this land to grow American pines and to concentrate hoop and bunya pine on the rainforest soils in the Brisbane and Mary Valleys. In addition, the fast growing American pines would also be planted on coastal lowlands as close to Brisbane as possible, as that was the market centre.

Meanwhile, forest pests were becoming a major problem in the existing plantations. Wallabies and rats were causing severe damage to young trees and the survival figures were getting worse all the time. Wire netting fences were erected around the plantations and traps of various kinds were used, but they did little more than contain the problem.

Weeds, too, remained a serious problem. Inkweed, peach, wattle and bellbush were the most troublesome in the early stages of plantation development and these were later joined by lantana. Maize was eventually used to stifle weed growth. Maize seeds were hoed into the ground, usually before the tubed tree saplings were planted. The crop of maize was used to feed the station's stock animals and the rest was sold.

During the 1930s the early silvicultural techniques were progressively refined in the light of experience and intense research.

Maize was no longer grown and weeds were removed by chipping or pulling. This was laborious but effective, especially if done intensively in the first year of a newly established plantation. It was, however, much more difficult in older plantations because by then lantana had become established and its rampant growth was difficult to control.

This period also saw the introduction of cold storage of hoop pine seed. Natural seed lasted only about twelve months but experiments now showed that seed held in cold storage lost none of its germination ability after four years. The seed still had to be collected from mature trees, and seed was produced only in occasional years, but at least this method meant that there was now a continuous supply for the nurseries.

In 1935 pruning of young trees was introduced, almost certainly for the first time. By removing the lower branches the tree went on to produce timber that was clear and free of knots. Thinning had also started in the older plantations but this was still on a small scale. Also at this time the earlier plantations of kauri pine started to go into decline. This was caused by thrips and although infestations were sprayed this did little to arrest the decline.

This was a good example of the difficulties that always lurk in forestry. Until an experiment has reached maturity, which might be measured in decades, there can be no certainty that it is a success. Problems of insect attack or soil properties may not affect a tree until it is in the last few years of its useful life. Invisible until then, they can render the whole exercise futile.

By the end of the 1930s nearly 8000 hectares of softwood plantations had been planted with stock raised at nineteen nurseries. Native pines now accounted for 80 per cent of the annual plantings of softwood, and of these by far the majority were hoop.

Meanwhile, the use of exotic pines was also being developed, although not without some difficulty.

The first planting of American pines had been made at Beerwah in 1924 with seed obtained from Florida, which had been selected because its summer rainfall was almost identical to that of southern Queensland. Most of the stock was slash pine. Further plantings were made two years later on 6000 hectares near Brisbane and additional species were introduced.

By this time some 600 hectares of native softwood plantations had already been established and the same techniques were now applied to the development of exotic plantations. This meant the use of shaded nursery beds and tubed stock.

However, the unreliability of winter rains which made tubed planting desirable with hoop pine proved to be less of a problem with exotics. Slash and loblolly pines are dormant in the winter and are thus able to handle harsh winter conditions more readily than hoop. By 1929 it was clear that shaded nurseries were not required for exotics and the plants did not need to be tubed.

In 1932 extensive plantings of slash pine were made at Beerwah and the early results were encouraging. The concern over winter rain, although still important, was not as strong as before and open root planting was now the accepted method. In 1933 at Beerwah: 'Planting was considerably delayed awaiting the winter rains. After a fall of 70 points planting was commenced only to be followed by a further dry spell. Results, however, have been particularly good and the care taken in planting has been fully justified.'

By the early 1930s slash pine and loblolly had emerged as the best of the exotics that had been used so far and they were heavily used in the new plantations in the wallum districts.

And it was at that point, just as everything seemed to be going so well, that disaster struck. Quite suddenly the exotic pines started to show signs of distress. Their needles twisted together and stuck to each other, shoot growth was depressed and the trees became shrubby and stunted.

It became obvious on a wide scale in 1934 and it was called fused needle disease, even though the needles were only a symptom of a more serious and debilitating condition within the whole of the tree.

The occurrence of fused needle disease varied considerably. In total only about 6 per cent of the exotic plantations were affected, although this was not known at that time. What was far more obvious was that in some stands it destroyed nearly 80 per cent of the trees. The immediate problem was to find out what it was and then develop a cure or some form of protection. Exotic plantations were an

GROWING UP

Lifting slash pine plants at the nursery at Beerburrum in 1970. These plants were used in plantations at Beerburrum and Beerwah.

essential part of the forest strategy in Queensland and this disease could make it all unravel.

In 1935 it was established that the disease was not a virus but it was three years later before the problem had been identified:

> The investigations concerned with the 'fused needle' in exotic conifers have advanced considerably and it appears that the diseased condition is intimately bound up with microbiological relationships in the soil. The application of phosphate dressings in affected areas has resulted in a change in the direction of these activities with consequent beneficial effects towards the pine trees. Trials with fertilizers other than those containing phosphorous have given no beneficial results. Zinc sprays and root grafts have also given negative evidence.

So it was proved that fused needle disease was not a disease at all. It was the result of phosphate deficiency in the soil and this was quite easy to remedy. Phosphate is still used in exotic pine plantations and fused needle is now a thing of the past.

Another exotic pine, the Caribbean pine from Honduras, was introduced in 1947. Trial plantings were made at Beerwah and Tuan Creek near Maryborough and these showed that this strain did not have a winter dormant period. Because of this it could not be planted open rooted and losses were heavy. Nevertheless it was a very promising strain as it grew well and produced superior wood.

It was not until 1976 that a method was developed that enabled

A demonstration of traditional skills at the Wood Works Museum at Gympie.

Hardwood forest at Mount Mee, north-west of Brisbane.

View of State Forests from one of the fire towers at Imbil.

The breaking-down saw at Guest's sawmill at Dirranbandi.

One of the control consoles at the fully automated sawmill built by Hyne & Son at Tuan near Maryborough in the late 1980s. Photograph courtesy of Hyne & Son.

PLANTATIONS

full scale commercial planting of Caribbean pine, and this has been described as one of the department's greatest achievements.

To promote the survival of Caribbean pine after open root planting a technique was introduced whereby the seedling was hardened off before it was planted. This was done by frequent root pruning and topping and the roots were dipped in clay after lifting. This, together with the culling of unsuitable seedlings, greatly increased the survival rate.

Queensland is the only place in the world where this technique is used although Caribbean pine is an important plantation tree in more than thirty countries. It is a prime example of how original techniques need to be developed in order to grow exotics in regions that are strange to them.

This work was certainly worthwhile as Caribbean pine planted on well-drained sites has a volume 20 to 40 per cent greater than slash pine and also has superior wood qualities. Thus the size of the timber resource in Queensland increased as soon as Caribbean pine was commercially planted.

Although this was a dramatic step forward, most silvicultural techniques in plantations changed very little from the original experimental plantings until the early 1960s. What changes there were were refinements rather than changes in principles.

Site preparation was done by hand and consisted of felling and brush hooking to clear the site. The site was then burnt and any unburnt material was laboriously reheaped and burnt by hand. Planting was then done by hand. This was followed by weed control, which was also done by hand, and the plantation was pruned and thinned using manual saws and horses for snigging. Fire protection has always been important in plantations as these species have little resistance to fire. This protection consisted of green scrub or hardwood stands surrounding the plantation area or, in the case of exotics, cleared firebreaks around and through the plantation.

In the 1980s establishing new plantations was a much more sophisticated process which passed through several stages.

The proposed site had to allow for the location of firebreaks, roads and tracks to provide good fire protection as well as access for management and logging. The plan also showed areas that were not to be cleared and these included creeks, wildlife corridors and other areas of special significance. In exotic plantations in particular the soil was studied in great detail and predictions made about the flow of surface water so that erosion could be kept to a minimum.

When this had been done, the merchantable timber on the site was logged and the area cleared by machine and then burnt. That

A stand of select hoop pines in a Queensland plantation, 1969.

was about the only preparation needed for hoop pine but with exotic pines the site might then be strip cultivated or mounded in areas of poor drainage which had been identified in the survey. Today, native forest is not being cleared for plantations. Instead the majority of new planting is on sites that have already produced their first plantation crop.

The planting rows are treated with herbicides to control weed growth and in a hoop plantation a cover crop such as oats or kikuyu is sown between the rows of planted trees. Hoop pine is planted as tubed stock in early summer at the rate of 830 stems per hectare. Exotic pines are planted as either tubed stock or open root in the summer in central and northern Queensland and in winter in the south-east. Density varies depending on whether the plantation is being grown entirely for sawlogs or for sawlogs and pulpwood.

Hoop pine plantations are always established on fertile sites and no fertilising is necessary. Exotics, however, need to be fertilised with phosphate if earlier studies have shown this to be deficient on the site. This is done at the time of planting, either from the air or along the rows on the ground. Competition from weeds is controlled by herbicides or by mechanical slashing until the trees are big enough to look after themselves.

Thinning is carried out when the trees are about three metres high. This is done to remove poor quality stems and to increase the growth rate of the remainder. Further thinning is carried out later and as the trees are now substantial these thinnings can be used commercially.

Pruning is done in two stages for both native and exotic pines, although the stages vary between the two. In all cases, however, pruning is done to a height of about 5.4 metres, after which the tree grows naturally.

Mature plantations are harvested by clear felling and site preparation for the next rotation starts immediately. By this time a great deal is known about the site and the next planting will be in the ground within weeks or months of the previous rotation being harvested.

All this work relies heavily on the ability of nurseries to produce adequate quantities of seedlings of the desired species, which in turn depends on the availability of seed. The department produces more than half a million hoop pine seedlings and over five million exotic seedlings each year. While most of these are used to stock plantations, the surplus is also used to reforest water catchment areas in local shires and other types of land care.

Today, most new houses in Queensland are built with frames made from plantation timber. Indeed, many plantations and their local privately owned sawmills are almost entirely dedicated to producing timber for this purpose.

Plantations are now the major source of timber supplies in Queensland and that is only possible because of the foresight of Jolly and, especially, Swain. Without their perception and their courage in committing the department to this essential activity, Queensland would today be importing nearly all its timber needs, and at considerable cost.

Training the Foresters

The need for trained forestry staff was apparent from the earliest years of the department. As far back as 1911 the first Interstate Forestry Conference passed a resolution supporting a forestry school which would take students from all states. It was, however, a long time before that came to pass. The difficulty was that while all heads of departments wanted trained foresters, they wanted them to be trained according to their own philosophy and to their regional requirements. The gulf between those who favoured European forestry techniques and those who looked to America was too wide to be easily bridged and seriously frustrated early attempts to introduce formal forestry training.

The first forestry school in Australia was established in Adelaide in 1911 and by default it became the 'national' school even though it paid little attention to the needs of other states. This school operated until 1925 and it was followed by the Australian Forestry School which opened in Canberra in 1927. This school was too Eurocentric for Swain's liking, but as a national school he had no choice but to send students there.

In 1965 this school was transferred to the Australian National University at Canberra, where it still operates as the Department of Forestry. Students follow a four-year course and graduate with a BSc in forestry. An honours degree is available after a fifth year of study. A degree course in forestry is also available in Victoria.

Graduates enter forestry as 'foresters', which is a technical rating open only to those with a degree. They are at the base of the higher echelons of forestry and can be expected to fill senior positions in due course. At the level of forester and above, all have this degree and many have further qualifications as well. Some people enter forestry as zoologists, botanists, economists and other disciplines and they have qualifications appropriate to their speciality.

There are a number of 'ranks' below that of forester and most require some kind of formal training, although obviously not to degree standard. This training has changed as the work became more sophisticated and forestry techniques improved. Indeed, the introduction of these techniques was often hampered by the lack of suitably trained staff to implement them in the field.

These training schemes changed as needs changed, and some were withdrawn and replaced by new schemes. The following describes some of the major initiatives used in training staff in Queensland forestry.

Forest Measurer Scheme 1930–1970

This was the first scheme to provide technical training of a formal nature. Initially the programme ran for nineteen weeks and provided basic training in most aspects of practical forestry. However, it did not provide the wider knowledge needed by those working in northern rainforests and a special two-year course was introduced in 1958 to meet this need. The two courses existed side by side but not successfully because people who had completed the two-year course had no career advantage over those who had done only nineteen weeks. After 1961 the nineteen-week course was used only when there were severe staff shortages.

Forest Learner Scheme 1934–1959

This was a five-year apprenticeship open to men between sixteen and eighteen years old. Students received no formal instruction but instead learnt by being involved in forestry activities in the field. Each year was spent in a different branch, starting usually with a nursery and finishing in the fifth year with administration and supervising others. In 1946 this scheme was extended to provide training in most forest types so that those completing the apprenticeship were no longer limited to types of their own region.

Forest Trainee Scheme 1959–1978

This scheme replaced the Forest Learner Scheme. Entrants had to have ten years' schooling and passes in English and advanced mathematics. It was open to men between fifteen and nineteen years old.

The four-year programme was more comprehensive than the scheme it replaced and students sat for formal examinations. In 1969 the scheme was reduced to three years and formal instruction was given in six 'schools' conducted during the three years.

Adult Trainee Scheme 1969–1977

This was an abridged version of the Forest Trainee Scheme and was designed as a quicker method of remedying staff shortages. It was open to men aged between twenty-one and forty-five with at least three years experience in forest industries. The programme lasted eighteen months and students attended four of the six schools of the trainee scheme.

Forestry Technical Assistant Scheme 1962–1967

This scheme was introduced to prepare people for careers in forest research or wood products research, neither of which were catered for by the existing schemes. The scheme demanded formal instruction for a period of seven years, but it did not provide it. The student was expected to attend specified external evening courses provided by academic institutions, although some could be done by correspondence. Entrants needed school passes in English and advanced mathematics and had to have completed the Forest Trainee Scheme. This scheme was not successful because of the high entry requirements and the length of the course.

Rural Technician (Forestry) Certificate 1968–1982

This grew out of the unsuccessful Forestry Technical Assistant Scheme when the Queensland Department of Education provided a correspondence course to replace it. The entry requirements were still high—passes in English, advanced mathematics and a science subject—but the student could do the course while still taking part in the forest trainee scheme, which meant it could be completed in five years. Students had to attend a two-week school each year at the Queensland Agricultural College and had to pass a written examination in all subjects as well as carrying out assignments.

Fellowship Certificate in Forestry 1979–1983

This was a thirty-month course taught at the Forestry Training Centre at Gympie and awarded by TAFE.

Associate Diploma of Applied Science (Forestry) 1979–

This is now the standard qualification required for most people working as overseers or rangers. It is a formal academic course requiring two years of internal study. Some units can be done by correspondence, in which case the course will take three years. The course is offered by the University of Queensland's Gatton College and the first two semesters are conducted at its campus at Gatton or by correspondence. When students have passed all the subjects in these semesters they enrol at the Forestry Training Centre at Gympie for internal instruction for the third and fourth semesters. In addition, students must complete sixty days work experience. This course meets the needs of both field and technical research functions and is accepted by the Department of Forestry at ANU as the equivalent of the first year of their degree course.

It follows that people entering forestry in Queensland as a career will either complete the Associate Diploma in order to work as an overseer and perhaps later as a ranger, or they will graduate from ANU or Melbourne University with a BSc(For.) and enter the department as a forester.

10 Fire

There are two types of fires in forests. There are those which have been lit by foresters and those which have not. There is a world of difference between the two.

Why on earth would foresters even think of lighting fires in a forest? All Australians are familiar with the dangers of bushfires and even though most have a healthy disregard for authority they would not hesitate to condemn anybody found lighting one.

Nearly every summer television shows dreadful pictures of entire hillsides burning. Houses are destroyed, sometimes people too, and vast columns of smoke sweep across nearby towns and cities. These fires, often starting from very humble origins, can spread at an alarming rate and change direction just as they seem to be under control.

The consequence of a forest fire is so obvious that one might think that forests and fires should be kept well apart. And for many years this was the view of foresters too. The policy was one of complete exclusion of fire from the forests. Unfortunately, though, forests produce a great deal of litter on the forest floor: dead leaves and branches, undergrowth that has dried and withered, strips of bark and even entire trees that have died and fallen. This litter can accumulate rapidly and in some places it might be a metre deep. When it is dry it burns very easily and rapidly and provides more than enough fuel to support a raging wildfire.

One of the many fire towers used for detecting fires in State Forests. This is the Gallangowan fire tower.

Excluding fire from the forests allows this fuel to accumulate as, apart from the natural process of decay, there is little to reduce it. The fire hazard therefore increases as the amount of fuel increases and if a fire does start it will be far more serious than if there had been less fuel to support it.

The policy of fire exclusion gave way, then, to a policy of prescribed burning. Prescribed burning means using low intensity fire in a strictly controlled way to reduce the amount of fuel on the forest floor and thus make a subsequent wildfire easier to control and less damaging.

This policy and its associated techniques developed slowly and, in forestry terms, relatively recently.

Until 1927 there was no control of any kind over rural fires in Queensland. But the drought of the previous year saw a series of major bushfires in the spring which threatened native forests as well as the newly established plantations.

The department supported an inquiry that was called to examine the problem of rural fires and this resulted in the *Rural Fires Act*, which was passed in 1927. Later legislation defined further penalties for the misuse of fire and set up a system of fire protection in rural regions.

Although the forestry department's policy in Queensland was the complete exclusion of fire that did not mean that it took no steps to defend the forests. Protection consisted of various forms of firebreaks. These included rainforest barriers which were naturally fireproof around hoop pine plantations, green breaks in cypress pine and hardwood forests, and cleared, cultivated breaks in exotic plantations.

These methods were in general use in forestry services throughout Australia and reflected the lack of knowledge of the effect of fire on different types of forests. They were also influenced by the severe damage that had been caused when fire had been used recklessly to clear land.

Major forest fires occurred in Queensland in 1951 and again in 1957 and these led to an awareness of the dangers of fuel build-up that resulted from complete fire exclusion. Prescribed burning was slowly introduced as a means of reducing this hazard but it was not until 1966 that it was used on a wide scale. There was a severe fire season in 1968 and damage in areas that had been treated with prescribed burning was far lower than it otherwise would have been.

Prescribed burning had been tested and proved and it was now widely adopted. As techniques were developed the use of prescribed burning was extended to potentially difficult forest types. In 1973 it was used operationally in slash pine plantations, burning of buffer strips was introduced to cypress pine forests in 1975 and in 1976 it was used on stands of radiata and patula to a limited degree.

The decision to change from fire exclusion to prescribed burning came about for a number of reasons.

In 1968 a forester was appointed to work full time on fire research and the results of this work, together with earlier experiments, showed that fire of low intensity caused very little damage to standing trees but it reduced the amount of fuel as well as assisting in weed control.

By the 1970s the cost of maintaining the earlier protection system of roads and firebreaks could no longer be supported. Prescribed burning offered an effective, if not better, protection for about a tenth of the cost. Further, the intense development of plantations, and the investment they represented, meant that far more protection was needed than had been the case with native forests.

The policy of multiple use of forests had also increased the fire danger. Firebreaks protecting native forests were primarily designed to stop fire sweeping into the forest from neighbouring land. While this had always been a danger because of lightning and accidents, the risks increased considerably when the public was invited to use the forests for recreational purposes. By reducing the fuel content of the forest, prescribed burning was much more effective in reducing the danger of fires that were started within the forest.

As with many forestry operations, the techniques of prescribed burning developed by a combination of trial and error and dedicated research and the techniques were changed in the light of new information.

The first method, which was used until 1971, consisted of lighting a continuous strip of fire around the perimeter of the block. The fire was usually lit in the evening and allowed to go out during the night. But sometimes it did not go out. A weather change overnight could produce a large fire front within the block itself which was difficult to fight and could result in considerable loss.

In 1971 strip lighting was discontinued and a method called grid ignition was introduced. With this method, the ground was divided into a grid pattern consisting (usually) of 40-metre squares. The intersections of these grid lines were lit by ground crews. The small fires they started spread slowly until they joined up with the next one and because each fire had consumed its own area of fuel, the fires eventually went out when they met.

The aim was to complete the burn within half a day. This required much expertise on the part of the field supervisor, who had to calculate the moisture content of the fuel (which determined the rate at which it would burn) and the effects of changes in the weather.

Lighting on the ground had limitations. The ground crews had to walk through the block and this was often slow and laborious. The amount they could burn in half a day varied from 10 hectares to about 200 and this was not enough to keep up with the protection programme. This meant that some areas were not burnt and the protection was therefore less than satisfactory. From 1975 fixed wing aircraft were used to light the grid from the air.

Aerial ignition allowed a much larger area to be burnt in the same time. The fires were lit by incendiary capsules dropped from the aircraft. The capsules were plastic containers filled with a measured amount of potassium permanganate and a machine in the aircraft injected each container with a liquid which spontaneously ignited the capsule about thirty-seven seconds later. As soon as they had been injected the capsules were dropped from the aircraft as it

Arming incendiary capsules in flight. This aircraft was igniting a controlled burn in the Maryborough District in 1972.

made a 'bombing run' and were safely on the ground by the time they ignited. The capsules then burnt intensely for several seconds and ignited the surrounding fuel.

The use of aircraft increased the area that could be burnt in a day to about 3000 hectares, and it rarely fell below 200. There were two major problems, however. One was the distance needed by the aircraft to turn around between each run, and the other was the large number of fires caused by capsules being dropped outside the block. This was not as careless as it might sound. The aircraft was moving rapidly even when at its slowest flying speed and a small error, perhaps because visibility was reduced by smoke, could lead to a significant error on the ground. It soon became clear that helicopters were much better for aerial ignition than fixed wing aircraft and they came into use about 1981.

Helicopters brought huge benefits. They could fly more slowly, they had better visibility and they virtually eliminated the problem of accidental fires. They could also turn much more quickly between runs and were able to light a block of 500 hectares in a single flight. They could also be used to ignite blocks that were inaccessible by foot or too mountainous for the safe use of fixed wing aircraft.

The helicopter remains the most efficient method of lighting large blocks and is now widely used even though it is expensive. Some smaller blocks are still lit by ground crews, however, where the cost cannot be justified.

Today, the use of prescribed burning is determined by the forest type.

Broadscale prescribed burning is carried out in all managed dry sclerophyll forests except where it might result in damage to regeneration. Burning is normally done on a cycle of three to four years and might also be carried out before logging. When the logging is finished the debris will also be burnt and then the block will revert to the normal burning cycle.

In wet sclerophyll forests burning is restricted to those sections which can be burnt safely during winter. Burning is rarely done in spring because of the risk of escapes and relights later in the season. Prescribed burning in these forests depends heavily on seasonal conditions but is rarely less than a five-year cycle.

Cypress pine is intolerant of fire but has the advantage of not producing much debris on the forest floor. Routine burning is therefore not carried out in this forest type.

In slash pine plantations burning is used to establish a mosaic pattern but this is not done until the trees are ten years old. If they are younger than that, their size makes them susceptible to crown scorch. The fuel conditions in plantations are usually more uniform than in native forests and they can be burnt with considerable precision.

Hoop pine, on the other hand, is very fire sensitive and will not survive even the mildest of fires. Hoop pine plantations are never burnt and the original technique of total fire exclusion is their only defence.

Although the techniques of prescribed burning have been refined over the years, it is certainly not a casual exercise. On the contrary, it requires much preparation, good judgment, perfect organisation and tight control.

The determining factors are the state of the fuel and the weather. There are only a limited number of days in each burning season that are suitable for prescribed burning and they have to be chosen carefully. If the fuel is too moist it will not burn well, and if it is too dry it can burn out of control. The moisture content of the fuel depends on how long it has been since it last rained. Samples of the fuel are taken and the moisture content is determined by the use of tables.

The weather is studied closely and accurate forecasts are obtained. Temperature, wind speed and humidity are particularly important because they can significantly reduce the moisture content of the fuel even while the burn is proceeding. In that case the burn can easily escape and do much damage.

If all the conditions seem favourable, test fires are lit and their behaviour studied. If calculations and predictions have provided incorrect information, which is rare, the test fires will make this obvious. The project is abandoned if the conditions are not exactly as they should be.

If all goes well, the required area will be burnt within a few hours and the fire will have extinguished itself. But it will be watched closely every minute of the operation. At the first sign of something going wrong defensive measures will be put into place immediately. Even if the operation produces a copy-book burn, as it usually does, there is considerable anxiety until it is successfully completed.

With a good burn there is very little damage to standing trees but there are certainly changes to the forest floor. Some species of shrubs, for example, are likely to respond vigorously after a prescribed burn, while others might be heavily reduced. Fauna is unlikely to be destroyed by the burn because of its low speed and low intensity but it will certainly be moved and its habitat possibly destroyed. These undesirable effects are reduced by burning the forest in a mosaic pattern in which burnt areas are interspersed with unburnt areas which offer refuge to those needing it.

In any case, the effects of prescribed burning are certainly not as dramatic as those from a wildfire, which might totally destroy a huge area of forest. The effects of prescribed burning are continually monitored in research programmes. Techniques are under continual scrutiny and changes are made as soon as they are shown to be necessary. The economic benefits of prescribed burning are more clearly defined. Not only are the effects of wildfires considerably reduced but the technique is also much cheaper than earlier methods using firebreaks.

The extensive development of plantations represents a huge capital cost and the timber they contain becomes more valuable the closer it is to maturity. That timber has to be protected during the whole of its growing life, perhaps thirty or forty years, and fire is its greatest enemy.

Fires that are not lit by foresters are much more damaging than

FIRE

Shortly after the capsule reaches the ground it ignites spontaneously for a brief period and creates a low-level fire on the forest floor.

those that are. Wildfires, whether lit maliciously or by lightning strikes, can do untold damage and lead to the loss of life and property. They are the forester's worst nightmare.

Queensland is fortunate in that it rarely experiences the massive conflagrations that occur from time to time in the southern states. There, dry conditions combined with high temperatures, strong winds and low humidity lead to extremely hazardous fire conditions and a single spark can produce a fire that might burn for days across many kilometres.

Those conditions are not so common in Queensland. The state rarely dries out for very long, especially in coastal areas which support the bulk of the forests, and extremes of temperature and humidity are less frequent. The fires that do occur are rarely as intense nor burn for as long as they do in the south.

Nevertheless Queensland often presents a very serious fire risk in certain weather conditions. In the south-east of the state, for instance, the worst fire weather is usually produced by a low pressure system to the south or south-east of the continent with a trough extending into Queensland. A cold front with thunderstorms followed by a westerly change can then make the region extremely vulnerable to fire. Rain associated with the thunder will often put out fires caused by lightning strikes, but the dry westerly winds after the rain rapidly increase the fire danger level and many serious fires have broken out in these circumstances.

Any fire needs three elements, all of which must be present to sustain it. These are fuel, oxygen and heat. If any one of them can be removed, the fire will go out. In a forest fuel will always be present in the litter on the floor and might be there in large quantities if it has not been reduced by prescribed burning. The trees themselves are not regarded as fuel, although they will certainly become so if the intensity of the fire increases. Of the other elements, oxygen is in the air and heat comes from whatever ignited the fire, and then from the fire itself.

The techniques of fighting a fire are directed at removing at least one of these three elements. In the forest there is a different method for each of the three.

Fuel can be removed by cutting a clear break between the rest of the fuel supply and the advancing fire. Oxygen can be removed by covering the fire with soil, and heat can be removed by applying water to the lower part of the flames. The difficulty is that whatever technique is used in a forest fire usually has to be carried out on a large scale and often in country which is difficult to access. In practice the three techniques might be reduced to a choice of only one, and that is likely to be the removal of fuel.

Removing fuel is done by making a break ahead of the fire. This means clearing a strip of ground until it is down to bare earth. The location of this line has to be chosen carefully as the break must be completed by the time the fire reaches it or all the work will have been wasted and personnel will be at risk. Natural features such as creeks and tracks are used wherever possible.

The line can be cleared by a small gang working with rake hoes. They work side by side and in the same direction, each clearing a part of the line. When one person reaches newly cleared ground that has been worked by his neighbour he calls for the gang to 'step up'. Each person then moves along the line until they reach the next part of uncleared ground, where they start work again. This means they never have to pass each other, which can be dangerous when visibility is reduced by smoke. The last person in the gang makes sure that the line has been properly cleared.

Although working by hand might seem slow, an experienced and fit gang can clear a considerable length of line in a relatively short time. It is tiring work, though, and progress drops off as time goes on. A bigger line can be made much faster by a machine such as a dozer or grader and they are always used when they are available. A

A bushfire on Fraser Island in October 1968.

clearing gang is then used only on parts of the line which might be inaccessible to the machine. Machines also need skilled operators and can be a hazard to others in poor visibility.

Although a firebreak will stop the fire by depriving it of fuel, it will not stop burning debris being blown across it to ignite fuel on the other side of the break. This is the biggest problem with this method. 'Spotting' will start new fires behind the break and render it useless as well as threatening the safety of the gang.

A development of this technique is called back burning. The fuel cleared from the break is pushed to the side nearest the approaching fire and when the break is complete this fuel is deliberately lit. The fire burns towards the advancing line of fire, destroying the fuel as it does so. By the time the two fires meet there is no fuel left to sustain either of them.

Spotting is again the problem. Burning embers can be lifted high on the convection currents and then carried along by the upper wind. If the wind is strong these embers can be carried for kilometres and still be burning when they reach the ground. They are not easy to detect in rugged terrain and might be difficult to reach even when they have been identified. The result will be a new fire front that is difficult to control.

The back burn might also take a different course to that

intended, perhaps because of a wind change, in which case it will become part of the wildfire rather than the solution.

Aircraft are used principally for detecting spot fires and to provide accurate information about the progress of the main fire. Fire retardants and water dropped from aircraft have proved unsuccessful in most Australian forests although they are widely used elsewhere.

The strategy used in fighting the fire is determined by the fire boss, who is in charge of the whole operation. There are three options: a direct attack, an indirect attack, and a combination of the two.

In a direct attack the effort is aimed at the front of the fire itself, which becomes the control line. Usually this is done only against small fires when the fire danger conditions and amounts of fuel are relatively low. Smoke and heat can make work difficult and the chances of the fire spotting across the line are high. On the other hand, the fire has little chance of picking up speed and the amount of burnt ground will be kept to a minimum.

Indirect attack uses back burning from a control line that is established well ahead of the fire, perhaps even several kilometres away. Back burning is an essential element of this strategy and is always used where there is fuel between the approaching fire and the control line. It is the only method which is effective against large or fast moving fires.

The fire boss has many choices when deciding where to establish the control line, and this is one of the advantages of this strategy. The line can be located to take advantage of natural features and can be far enough away from the advancing fire to allow time to construct the control line properly. It will be less successful, though, if the back burn fails to burn all the intervening fuel and in any case it establishes a solid area of fire which can get away. The burnt out area is also considerably increased.

Both strategies are used at the same time when the head of the fire is too hot or moving too quickly for a direct attack. The fire is then controlled by a direct attack on the flanks of the fire to contain its spread, while an indirect attack is being mounted against the head.

Attacking the flanks is a standard method in nearly all fires because it reduces the risk of the fire suddenly developing a new head along one flank. It is usually quite impossible to mount a direct attack on the head of a forest fire until late in the day, when wind and moisture content usually swing in favour of the fire fighters.

The sooner the fire can be attacked the more chance there is of bringing it under control before it has done much damage. Once the fire starts to spread the job of controlling it becomes infinitely more difficult.

In periods of high fire danger forest staff are rostered on fire duty outside normal working hours so that a fire crew is available at all times. Fire towers are staffed and a state of readiness is maintained so that the reaction to any outbreak can be as quick as humanly possible.

In anything but the smallest fire the fire boss has the most difficult job of all. He has to make the strategic decisions, knowing all the time that if he gets them wrong he might endanger the lives of his crew or burn out many hectares of valuable forest. He rarely has as much information about the progress of the fire as he needs and has to decide when to replace a crew with a fresher one, when they should eat and where, and maintain fuel supplies to the machines. All this will be done in circumstances which are clearly very dangerous.

People in the gangs work in even more trying conditions, often within easy sight of the fire and surrounded by smoke. Whatever their job, it has to be done carefully otherwise it will have no effect. This is not easy when the fire is close at hand and the instincts demand an urgent evacuation.

If the fire is a large one, the forest staff will be joined by the local rural fire brigade, local landowners and others who can be useful. This can sometimes lead to confusion about the command structure or other problems such as different radio frequencies and a lack of intimate knowledge of the layout of the forest. Nevertheless, qualified help is invariably welcomed as the number of people on the ground provides much more flexibility in strategy as well as providing relief for earlier crews.

When the fire is finally put out the crews start the laborious and tiring process of mopping up. This means examining the whole of the burnt area and making sure that there are no small fires burning which could flare up again. Burning embers and small fires are doused with water, burning logs are raked out and burning trees felled and extinguished. This work might take several days or even longer, but the job is not finished until it has been done.

One of the worst forest fires in Queensland occurred in Toolara State Forest on 22 September 1991 and it provides a good example of the reality of a forest fire.

The temperature that day was 31 degrees Celsius and the relative humidity was 12 per cent. There had been virtually no rain for months and the fire danger rating was very high. At 12.09 pm a

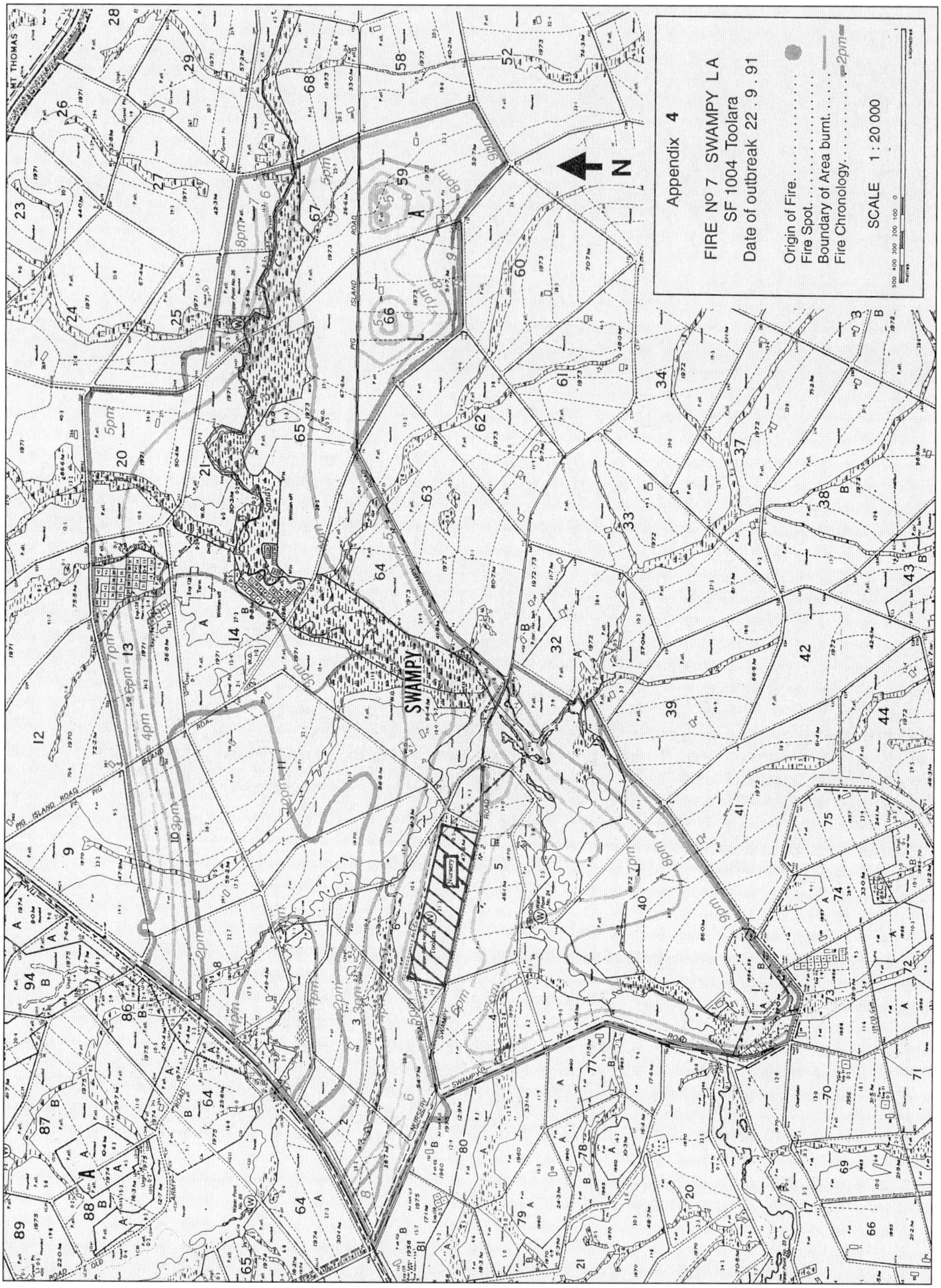
Appendix 4
FIRE Nº 7 SWAMPY L A
SF 1004 Toolara
Date of outbreak 22 . 9 . 91
Origin of Fire
Fire Spot
Boundary of Area burnt
Fire Chronology
SCALE 1 : 20 000
N
SWAMPY
PIG ISLAND ROAD
NURSERY ROAD
SWAMPY ROAD
MT THOMAS

FIRE

small fire started on the edge of the road from Gympie to Tin Can Bay which runs through the Toolara State Forest. The fire was thought to have started from sparks thrown out by a broken trailer being towed behind a bus that was taking members of a fishing club back to Brisbane.

Motorists following the bus stopped to try to put out the small fire and they were seen by a forest worker who was returning along the same road after dropping a colleague off at the Kelly fire tower. He was driving a four-wheel drive Toyota troop carrier which unfortunately could not carry a mop up unit. Had it done so the fire could probably have been checked at that early stage.

He radioed the information to Toolara Forestry Headquarters and then went to help the public firefighters in Sandy Creek. But a strong south-westerly breeze made the fire spread rapidly and he asked the volunteers to withdraw for their own safety.

When the news was received at Headquarters three men left quickly in two Land-cruisers. When they reached the fire they decided to try to hold it on a track between the pine plantation and the native forest buffer strip which ran along the creek. While driving down the track they saw spot fires in the plantation and stopped to put them out.

At about 12.20 a huge fire storm spread from the creek and engulfed one of the vehicles containing one man. He ran from the vehicle towards the road but he was severely burnt and was later transferred to the intensive care unit at a Brisbane hospital.

Meanwhile, the other two men in their vehicle tried to follow the track around the edge of the plantation but it came to a dead end at the steep side of a swamp. They turned around to drive back through the fire towards the road but the vehicle became caught on a stump. The men had to shelter in the vehicle for ten minutes while the fire passed over them. During this appalling time they sprayed themselves, inside the vehicle, with water from the mop up unit. Had they not done so they would almost certainly have been killed. When the tyres were well alight they decided to abandon the vehicle and make for the road. Both were burnt but did not need hospital treatment.

A D4 tractor reached the fire at 12.35 and started to build a fire line across Sandy Creek to stop the fire from spreading. This had to be abandoned because of the weather conditions and dangerously high fire rating. Attempts were also made to back burn along the western edge of three compartments but these were unsuccessful for the same reasons.

By now the fire was spotting up to a kilometre away and it was

Opposite: *This map shows the progress of the fire at Toolara State Forest on 22 September 1991. Notice how quickly the fire spread in the early stages, and the huge increase in the area of the fire between 3 pm and 4 pm.*

clear that it could not be stopped in the existing conditions with the men and machinery that were available. Efforts were therefore directed at the flanks of the fire in order to contain them. At about 5.30 the wind shifted to the north-east and although this brought the forward spread of the fire under control it also produced new problems. The attempted back burns now became forward burns and they raced through the swamp in the middle of an area that was being logged.

This dry swamp had not been burnt for twenty years and it was impossible to stop the fire jumping across the Swampy Road, which runs along the southern edge of the Toolara nursery. The nursery buildings now came under threat because back burning along the northern edge of the nursery had not gained sufficient depth to be effective. The buildings would certainly have been lost had not one of the staff turned on the irrigation sprinklers.

At 7.30 the wind changed again, this time to the south-east, and dropped in strength. This, together with an increase in humidity, allowed back burning to be completed from the major roads. These were successful and the fire was finally contained by 9 pm. All fire units then started to black out stumps, smouldering heaps and peat swamps. The fire produced a heavy fall of pine needles and these fuelled several small fires until rain fell on 19 October.

Nearly a thousand hectares of valuable plantation, which was about twenty years old, were severely burnt. In addition to this loss, the cost of fighting the fire came to over $67 000.

This fire illustrates some of the points made earlier in this chapter. Although a quick initial attack would probably have contained the fire at an early stage, this was not achieved and the resulting loss was very high. The fire soon became so fierce that effective attacks could not be mounted until the conditions changed and the fire danger rating reduced. Until then only the flanks could be contained. Wind changes altered the path of the fire dramatically and if this had not been met by a rapid response damage would have been considerably greater.

It is not surprising, in the light of this account, that fire is one of the greatest enemies of the forests. And fighting them is one of the most dangerous jobs that field staff can be called upon to do. Few other people would wish to do it for them.

11 Fraser Island

Fraser Island means different things to different people. It might mean fishing, camping, or one of the last examples of readily accessible rainforest—magnificent stands of tall timber that seem timeless.

But until the 1960s Fraser Island did not mean much to most people. Except to foresters and those in the timber industry. To them it was a source of some of the best wood in Queensland. They had lived on Fraser Island, worked on it and looked after it for nearly ninety years before most people even knew it was there.

Then all that changed. Suddenly everybody knew about Fraser Island, and they thought it was priceless.

Fraser Island is thought to be the biggest sand island in the world, with huge sand hills rising nearly 250 metres above sea level. These have been formed by the action of the wind, which has produced a peculiar system of hills, ridges and valleys. Sand has been blown into the steep sided valleys to engulf whatever grew there.

The landscape is not uniform throughout the island. In the middle the ground is higher and steeper and it is here, in the centre of the island, that one finds areas of rainforest. Surrounding this are forests of different types, which in turn give way to coastal heath. It seems amazing that such massive trees can grow on what is virtually pure sand. Indeed, the only nourishment comes from the decaying remains on the forest floor.

Felling a huge satinay tree on Fraser Island in 1922.

Dotted throughout the interior of the island are a number of lakes that still defy rational explanation. They are in the higher regions and, given that the whole of the island is sand, their presence is something of a mystery. But they have great appeal. The lakes are surrounded by timbered slopes whose vegetation often comes to the very edge of the water. They are placid, silent and good for the soul.

The eastern, seaward side of the island consists of sweeping beaches that run almost the entire length of the island. Wide and firm, the beaches serve as the main north–south road and even as landing strips for light aircraft.

The first European to see Fraser Island was Captain Cook, although he did not know that it was an island. In May 1770, while sailing up the east coast of Australia in the *Endeavour*, he saw and named Double Island Point and Wide Bay to the north. From the ship he saw a group of Aborigines on the shore and named the point Indian Head.

The Aborigines had a long wait before they were visited again. In 1779 Matthew Flinders arrived in the *Norfolk* and he sailed down the west coast, in Hervey Bay, and named several features as he went. He still did not know that the land was an island but he 'entertained a conjecture that the Head of Hervey's Bay might communicate with Wide Bay'. He was unable to prove this, however, as he was unable to take his ship into the Great Sandy Strait.

Flinders returned to the area in July 1802 in the *Investigator*, with the *Lady Nelson* serving as a tender. On 30 July three parties landed near Sandy Cape and they were the first Europeans to set foot on the island. A party of naturalists collected specimens, another party led by the captain of the *Lady Nelson* gathered wood for fuel, while Flinders led the third party to the northern tip of the island. He was not overimpressed by what he saw, although his description is as true now as it was then: 'This part of the coast is very barren; there being great patches of movable sand many acres in extent through which appeared in some places the green tops of grass, half buried, and in others the naked trunks of such as the sand has destroyed.'

Flinders was on a voyage north and had no time to spare for exploration, which was unfortunate as his uncanny skill for taking ships through waters which even modern sailors find tricky would surely have proved that this great sand mass was an island.

It was another twenty years before this was finally established. In 1822 Governor Brisbane sent Captain Edwardson north in the cutter *Snapper* to find a river suitable for a new penal settlement. Edwardson succeeded in sailing through the Great Sandy Strait and so proved that the sand mass was not attached to the mainland. That was about all he did achieve, however. He failed to locate the Mary River and instead thought that Tin Can Bay was a river. He sent a good report of the anchorage in the Great Sandy Strait but the convict settlement was established two years later at Moreton Bay instead.

The island was again forgotten for many years as it seemed to have little to offer. Then in 1836 there was a chain of events that not only gave the island its name, but made it famous throughout much of Europe.

During that year the brig *Stirling Castle* was making her way back to London in ballast, having delivered a cargo to Van Diemen's Land. On the night of 22 May 1836 she struck a reef far to the north of the island and was totally wrecked. On board were eighteen people, including the master, James Fraser, and his pregnant wife Eliza.

They managed, with some difficulty, to launch the ship's pinnace

Loading eucalypt logs on the tramline in 1929.

and a longboat and they set out for the nearest European settlement, which was at Moreton Bay about a thousand kilometres to the south. Initially the pinnace carrying Fraser and his wife was towed by the longboat manned by the crew, but the crew soon realised that their slow progress meant their survival was unlikely. One night the crew cut the pinnace adrift so that they could make better progress. The longboat eventually came ashore near what is now the Gold Coast and the crew started to walk south. Only one person survived.

Meanwhile those in the pinnace started a journey that was to last six weeks, during which Eliza gave birth to her child which survived only a few hours. On 26 June 1836 they came ashore near Waddy Point on the sandy island and were immediately surrounded by Aborigines. The men were taken away and Eliza was left to spend the night alone on the beach. In the morning she was taken to the camp, where she rejoined the others. From that time Eliza was expected to work with the Aboriginal women. She was, she said, regularly beaten and was rarely allowed into the huts at night even when it rained.

Some of the survivors decided to escape and make their way south. They eventually made their way to Bribie Island, where they met some settlers from Moreton Bay.

When the authorities in Brisbane learnt that there were survivors of the *Stirling Castle* and that some of them, including Eliza, were held captive by the Aborigines they immediately launched a rescue mission under the command of Lieutenant Otter.

With the help of two educated convicts, John Graham and David Bracefell, both of whom had lived with the Aborigines, Otter learnt that Eliza had been taken to a corroboree ground at Cooloola, while the other survivor, Baxter, was on the southern tip of the island. Graham took a canoe across to the island from Inskip Point and rescued Baxter, then returned and walked 50 kilometres down the Cooloola beach to rescue Eliza.

The captives had been held by the Aborigines for fifty-three days after they had landed at Waddy Point and during that time Captain Fraser had died. There were several accounts of his death, but the most likely is that he was speared when he was unable to carry a large log and that he died from this wound about a week later.

When Eliza Fraser returned to England in 1837 her story fired the imagination of the entire country and she turned it to good purpose. So much so that her account of her experience became more dramatic the more it was repeated. It was repeated frequently and very profitably and Fraser Island became famous in the process.

Apart from the fact that people in Queensland also now knew of Fraser Island, Eliza Fraser's experiences did not lead to European settlement of the island. Instead this came about in 1842 when the convict settlement at Moreton Bay was closed and the area opened up for settlement.

This attracted Andrew Petrie, who had once been Superintendent of Public Works in Brisbane. He set out with Henry Stuart Russell to explore the region and in doing so they met up with David Bracefell, who was now living with Aborigines near Noosa. Bracefell acted as their guide and eventually they camped near Snout Point on Fraser Island. Petrie's account did much to determine the future of the island: 'In this scrub I found a species of pine not known before. It is similar to the New Zealand cowrie pine and bears a cone. It forms valuable timber ... the formations and productions of the island are much the same as those of Moreton Island; the timber is a great deal superior, and also the soil; the cypress pine upon Fraser Island being quite splendid.'

In 1847 a town site was surveyed on the Wide Bay River and the following year a settlement was established in what is now part of Maryborough. When gold was discovered at Gympie in 1867 Maryborough became an important supply centre and this was helped by the fact that many ships en route to London from Sydney avoided

GROWING UP

McKenzie's jetty, Fraser Island, about 1926.

the slow detour up the Brisbane River and instead landed their Brisbane cargo at North White Cliffs on Fraser Island. They were met there by merchants from Maryborough. At about the same time a few selections were taken up on the island to support grazing animals taken across from Maryborough.

Andrew Petrie's account in 1842 of good timber on the island led to his son going there some twenty years later in the company of William Pettigrew to examine its prospects for timber.

Pettigrew had been born in Scotland in 1825 and was one of the younger sons of the Laird of Tarshaw. He chose a career in medicine but when he fainted at his first operation he decided that surveying might be a little less gory. When he was twenty-three he was recruited as a surveyor by the Reverend John Dunmore Lang, who was anxious to bring Scottish migrants to Queensland.

The party Lang raised, including Pettigrew, arrived in Brisbane in 1849 in the ship *Fortitude* and they named their first settlement the Fortitude Valley. Pettigrew managed to obtain work as a surveyor with the Crown Lands Commissioner. It was while in this position that he began to realise how suitable the trees were for timber and, from that, how profitable a sawmill would be. In 1852 he started the first steam-driven sawmill on the corner of William and Margaret Streets in Brisbane.

His investigation of Fraser Island with Tom Petrie soon confirmed that the timber there was abundant and of very good quality. He immediately started to build a sawmill in Maryborough about ten kilometres downstream from the wharf, and in 1863 the firm of W. Pettigrew and Co. cut its first log, a kauri pine that had been brought from Fraser Island by punt.

From then on development was fairly rapid. In 1866 Wilson, Hart built their sawmill as did Hyne & Son, who built a huge mill close to the town. Although a shipment of cedar had been sent to Maryborough from Bundaberg in 1867, all these mills were established to mill timber from Fraser Island. By 1869 timber-cutters were working full time cutting kauri from Woongoolba Creek. These logs were taken to the beach by bullocks and rafted across to the mills at Maryborough.

By 1879 logging had extended north to Yidney and from there to Woralli and Bowarrady, where the cut included hoop, kauri and beech. Satinay was also cut but was rejected by the mills because of the damage it caused to saws and its tendency to shrink and warp. Blackbutt, on the other hand, was found to be superb, but even this was overshadowed by tallowwood, a magnificent hardwood which was often found in very large stands of huge trees.

Surprisingly perhaps, the first attempts at regeneration were made long before there was a forestry department, and indeed long before most people saw the need for one. In 1882 the District Surveyor, McDowgall, organised the planting out of 28 000 young kauri pines in brushed lines through and under thick scrub. By 1884 about a hundred hectares had been underplanted but the young pines were unable to compete with the mature trees around them and the experiment failed.

In 1906, by which time there was a forestry department, Wilson Hart and Hyne & Son, now the two biggest sawmillers in Maryborough, combined to purchase logging rights to 1800 hectares on the island for a stumpage of fivepence per hundred super feet.

Their biggest problem was getting the timber to the mill. Not only had the logs to be taken across the water to Maryborough, but they also had to be taken from the scrub to the island's coast. The sandy terrain made haulage difficult, especially for the quantities they now intended to handle.

The answer came from a man called Edward Armitage. In 1874 he had worked for his brother who was then manager of Ramsay's sawmill near Maryborough. Together they had persuaded Ramsay to let them build a wooden tramway to move the mill's logs. Ramsay agreed and in spite of having to get engines cast and steel tyres made, Armitage had the tramway running successfully in six weeks.

Armitage now suggested to the two big millers that a railway on Fraser Island would solve most of their problems. They agreed and shortly afterwards let a contract for the haulage of sleepers for the line. The track, consisting of steel rails each weighing 12 kilograms, was laid by Armitage. It ran south-east from the mouth of Urang Creek for about 6 kilometres and then ran a further 5 kilometres into the Poyungan scrub. On it ran a steam locomotive called *Doris*.

The central part of Fraser Island was declared a forestry reserve in 1908 and the first forestry camp was established in 1913 on Bogimbah Creek by forest ranger Walter Petrie, the son of Tom Petrie. Petrie's main task, as determined by Jolly, was to experiment with the regeneration of eucalypts, the treatment of hoop and kauri pines and the acclimitisation of exotic pines.

Following the shifting of the railway, in 1916 the camp was moved to a new site near the mouth of Woongoolba Creek. A large bungalow was built and a nursery was set up to support trial plantings. Unfortunately the soil in this area was not very good and mosquitoes and sandflies made life harsh for everybody. So in February 1920 the station was moved yet again, this time to a site about 10 kilometres up Woongoolba Creek which came to be known as Central Station. In the same year Petrie was appointed Deputy State Forester for Queensland.

Activity on the island had increased considerably after the end of the First World War. In April 1918 H. McKenzie, a Sydney timber merchant, bought the timber rights to 4000 hectares and undertook to cut and saw 250 cubic metres a month for the next ten years.

McKenzie immediately started building the first, and only, sawmill on the island. This was built near the old quarantine station, about three kilometres inland from North White Cliffs. The sawn timber was taken to a jetty which was built on piles of rough hewn satinay.

From the mill a tramline ran inland up a steady climb to 720, so called because that was the time the tram passed there each morning, north of Lake McKenzie. From there the main line ran east towards Lake Wabby and an old timber camp where there was a steam-operated log hauler. This was built in 1924 by two American engineers and consisted of a huge drum which held a steel cable. The teamsters snigged the logs from the scrub and the cable was run out to them, often half a kilometre or more, and the logs were pulled the rest of the way by the winch.

McKenzie's sawmill and jetty were abandoned in 1925, by which time most of the island had been declared a State Forest. The department continued to operate the tramline until 1935.

Under the guidance of Swain, Grenning drew up a working plan for silviculture on the island. Its object was the liberation, regeneration and fire protection of the eucalypt areas; the improvement of satinay and brush box stands by ringbarking useless trees; the planting of poorly stocked areas with hoop and kauri pine and regeneration of cypress pine. There was to be no treatment of hoop pine areas, although they had been given liberation treatment during the war years. This programme, with variations, formed the basis of the work on Fraser Island for many years.

The building boom after the Second World War produced a shortage of timber that led to a reappraisal of those timbers that had been unpopular in the past. This, and the new technology of kiln-drying, now saw a great increase in the use of satinay and brush box. Satinay in particular was found to have important qualities that had not been suspected until now.

Satinay forests are unique to Fraser Island and Cooloola. Satinay grows in dense stands and produces a straight cylindrical trunk up to 30 metres long. The quality that suddenly brought it into prominence was that it was found to be resistant to marine borers. This invaluable property led to it being used as sidings in the Suez Canal and to build wharves at Tilbury Docks in London, as well as being put to many similar uses in other parts of the world.

In the 1950s working conditions on the island were still as hard as they had ever been. Work was not limited by daylight and it was not unusual for logs to be loaded during the night using light from a torch of burning satinay bark. There were still five bullock teams in use at Poyungan, but the early tramways had disappeared and their sleepers had either been used for firewood or had been laid on steep parts of the roads to provide better traction.

By the early 1970s the sawmills at Maryborough were taking about 40 per cent of their logs from Fraser Island. By this time the department had developed two methods of silviculture. One was based on cutting only those trees that had been marked by officers of the department, so that logging was very specific. This method not only produced merchantable logs, but was also used to thin the stands to promote better growth by the remainder. The other method consisted of clear felling followed by burning. This was used with stands of blackbutt because the species needs fire in order to regenerate and logging without burning would have led to the permanent destruction of blackbutt stands.

In 1977 the department entered into long term contracts with

Logging train on Fraser Island. It finally went out of service in 1935.

both Wilson Hart and Hyne & Son. These were to run for twenty years and included an option for a further twenty years by agreement. This meant that the sawmills could expand and introduce new technology secure in the knowledge that they had long-term access to the timber on Fraser Island to justify the large capital costs that the technology required.

This arrangement gave security to the timber industry and at the same time showed the department's confidence in being able to sustain productivity of good quality timber from the island for many decades. Its confidence extended even further as the techniques of forestry, especially sustained yield, were designed to maintain the productivity of forests in perpetuity.

But times were changing and the future of the timber industry on Fraser Island was by no means as secure as it seemed.

Since 1925, when most of the island had been declared a State Forest, the department had virtually managed the entire island. In the 1960s about 10 000 hectares were lost to town sites and other public uses but the island was still largely a State Forest. The department had itself designated twenty-three beauty spots, totalling about 3000 hectares, and these were left in their natural state. In

The last tree to be felled on Fraser Island, December 1991.

1971 the department, which was still responsible for National Parks, declared about 25 000 hectares in the northern part of the island as a National Park and two years later a further 9000 hectares were added. The area of state forest was then about 120 000 hectares and it produced a cut of about 22 000 cubic metres a year.

Apart from timber, the only other usable resource on the island was mineral sand. Applications for sand mining leases had been made as far back as 1949 and some had been approved by the government, but it was not until 1971 that sand mining got under way on a large scale.

By then the State Government had already decided to stop sand mining at nearby Cooloola and the start of this activity on Fraser Island aroused immediate opposition.

In 1974 the Federal Government passed the *Environment Protection (Impact of Proposals) Act* which gave it specific rights to protect the environment. Although the State Government still controlled activities such as sand mining, the Federal Government was the only authority able to issue export licences. This Act of 1974 allowed it to refuse to issue such licences if the project was deemed to be harmful to the environment. Export licences were essential to sand mining as all the major markets were overseas.

Following pressure from conservationists, in 1975 an environmental inquiry was set up under this Act. It was almost swamped by the number of submissions it received, some of which were only remotely connected to the issue in question. The report of the inquiry was presented in October 1976 and it recommended that with the exception of one specific location the export of minerals from the island should be prohibited. This was accepted by the Federal Government and the following December the export licences were withdrawn.

In its report the inquiry pointed out that it was not its task to inquire into forestry activities on the island except as they might be affected by sand mining. It came to the conclusion that such effects appeared to be negligible although they might become more prominent in the future.

Inevitably, though, the inquiry did consider the effects of forestry on the environment of the island, if only in passing. It said that it was impressed by the responsible attitude of the department and by its good working relationship with the sawmillers in Maryborough. It thought that the integrity of the island had not been adversely affected by the 'carefully controlled logging operations' and that only highly trained observers would recognise that the island had been logged for so long. It expressed some reservations about clear felling, burning and chemical treatment in thinning and thought they might be detrimental if they were allowed to continue.

If the department thought that was the end of the matter, they were wrong. Many conservationists saw the banning of sand mining on Fraser Island as merely the beginning, not the end, and mounted a strong campaign to prohibit any logging on the island.

In December 1989 the Australian Labor Party won the state election in Queensland and took over after many years of National–Country Party rule. The ALP had campaigned heavily on environmental issues and had promised an inquiry into the conservation and management of Fraser Island, which in practical terms meant an inquiry into logging. Within a few weeks of its election, on 26 February 1990, Premier Goss announced the appointment of a Commission to conduct this inquiry. The Commission, with Mr Fitzgerald as chairman, was formally appointed on 8 March.

The Commission received 544 submissions, including several volumes from the forestry department. All submissions were made available to those who had lodged one, and further submissions could be made in the light of them.

Many submissions were little more than emotional appeals, but others were highly technical statements written or supported by experts from a range of disciplines. In spite of the volume of this material, the Commission produced its report in May 1991, little more than a year after it started work.

The report itself was a work of some substance and those anxious to find its conclusions had to reach page 101 before they found recommendation number sixty. This said that the logging of blackbutt should be permitted in areas that had previously been logged, but apart from that 'logging of any timber other than blackbutt be prohibited'. Other than that concession it recommended that 'All State Forest in the region be revoked'.

Foresters and those in the timber industry were stunned by the decision and many thought that it was blatantly political—a debt that the new government had to pay to its conservationist supporters who, they thought, had helped it into office. They simply could not understand the reason for it. They had, they said, been logging Fraser Island for about a hundred years and in spite of that it was now so beautiful that it had to be protected. It seemed proof enough to them that they had hardly destroyed the place in the past and there was no reason to believe they would do so in the future.

As Mr Fitzgerald had suspected, logging only blackbutt was rejected as uneconomic by the Maryborough sawmillers and so, by a decision of the government, logging on Fraser Island came to an end on 31 December 1991.

Fraser Island subsequently succeeded in obtaining World Heritage Listing and it is now managed as a National Park. There is no forestry activity of any kind on the island and in most places it needs a skilled eye to see that there ever has been.

Aila Keto
Conservationist

Television reports of the activities of early conservationists generally showed them as a noisy group of people waving placards, shouting, obstructing machinery in the forests and confronting police, with the threat of violence never far away.

Those images helped to create a public perception that conservationists were demanding something for their own benefit, that their strident clamour was basically selfish, and that they did not understand the economic consequences of their demands. They were urban, they were excessive and they were noisy.

As one of the country's leading conservationists, Aila Keto is nothing like that. She is a quiet, professional scientist who believes strongly in applying scientific rigour to environmental issues and who abhors violence. She is articulate, she can assess complex arguments at the speed of light and she believes passionately in the need to conserve rainforests. There are those in Parliament and in the Forestry Department who would feel doubly blessed if they never heard of her again.

Aila Keto was born at El Arish near Tully in northern Queensland of Finnish parents. She went to Herberton High School and then to James Cook and Queensland Universities. She did an honours degree in biochemistry at the University of Queensland and taught there part time. She took a year off in Europe then returned to complete her PhD. When her son was born she took time off to enjoy parenting but maintained her scientific interests. Her background, then, is studying, teaching, scientific research, and parenting.

She has had a moral concern about conservationism since she was ten but did not become actively involved until the early 1970s when she helped organise a symposium at the University of Queensland which reinvigorated the Queensland Conservation Council.

While looking after her young son she read widely and saw the need for an organisation to facilitate community action to preserve Queensland's rainforests. There was then no organisation focused on this issue, so in 1982 she founded the Rainforest Conservation Society, which she still leads.

'My concern is about the fragility of the rainforest, which is an extraordinary ecosystem that has survived for tens of millions of years—from being very widespread to virtually just hanging on by the skin of its teeth. I think very few people recognise how fragile it is, particularly the relic aspects in the rainforest.'

Aila Keto was largely instrumental in having the wet tropical rainforests accepted for World Heritage listing, and in persuading the Fitzgerald Inquiry to ban logging on Fraser Island. Both issues put her in direct conflict with the forestry department and the National Party when it was in power.

One of the arguments she faced, especially in relation to Fraser Island, was that the area had been logged for about ninety years and here she was saying that it was so special that logging should be stopped.

> That was a pathetic argument. It does not relate to what World Heritage is about. World Heritage is recognising those aspects that are of outstanding universal significance to human kind, either from the point of view of science or conservation. There was no suggestion that Fraser Island had been logged to oblivion. But it *had* been logged. The forests had been changed structurally, the population structures had been changed and the original grandeur of the forests had been changed. But that does not mean to say that in a few hundred years it will not restore itself. So my answer is yes, it has been logged and it has been changed, but enough has survived to warrant the work and effort needed to bring it back.

She believes strongly in the need to maintain the strict requirements that must be met for World Heritage listing and that it should not be used simply because people, including politicians, want to stop the actions of others. She recognises that this can be a problem within the conservation movement. 'There will be ratbags in the conservation movement just as there are in industry. We have to guard against downgrading the concept of World Heritage, which is to ensure that we do not repeat the mistakes of past generations.' Her methods have little in common with some early conservationists. 'I have a personal aversion to violence or any form of aggression. I don't think you can ever solve things simply by direct action and I would never agree to it being used as a first resort.' But as a last resort? 'As a last resort I do agree.'

She has never arrived at the last resort.

Aila Keto relies on persuasion, persistence and communication. If people keep talking, then there will be a resolution. It might take time, and during that time a resolution might seem very unlikely, but at the end there will be one. And when it arrives it will be more acceptable to all parties because it has been reached without coercion.

Persistence is the key. She simply keeps on until the issues are clearly identified and the arguments assessed. She believes her arguments are superior (otherwise she would withdraw) and keeps with the issue until this superiority is accepted. Then change can follow.

Personalities do not come into this. 'I don't focus on people. I focus on the issues and the policies and I believe that one should be able to differ even amongst friends. If you criticise what a person is doing you are not implying that you don't respect that person's right to do it. There are people in the forestry department that I have enormous respect for, but I disagree with some of the things that are being done.'

She thinks the department has failed to move with the times. After doing much to preserve forests from alienation, the department, she says, became inward looking and comfortable. An example was the reluctance to bring in environmental guidelines as part of departmental policy.

'Scientists in the department knew what they should have been doing in the 1970s but they did not bring in official guidelines. Even when they did, in 1982, they were largely based on hydrological studies, which made them restricted. Their application involved checklists which were very cynically implemented.'

Had the approach been different, and had sensible guidelines been followed, she thinks the case for prohibiting logging in the wet tropics would have been much more difficult to promote than it was.

She is also critical of the influence that some sawmillers had on the National Party when it was in power in Queensland, and thinks that there was political interference in what the department was trying to do.

'As I apportion blame I think in order it would go first to the Coalition government and in later years the National Party government. And then the industry, because parts of it were self-seeking and politically motivated. And finally the forestry department which was caught in a crack where it did not have much leverage.'

With almost no violent confrontation, and with none of the excesses of early conservationists, Aila Keto has achieved change on a scale that many would have thought impossible.

Those opposing her changes no longer have to deal with an emotional, loosely organised band of campaigners. Instead, they face a scientific argument presented with conviction and persistence, and which will be defeated only by a better and equally rigorous argument.

Nothing personal, you understand?

12 Sawmilling

Sawmilling and forestry are two quite distinct and separate activities. One starts where the other finishes.

Forestry is concerned with the raw material, with growing trees that can meet the varied needs of the community. It is a form of agriculture and is both a science and an art.

Sawmilling is the first step in an industrial process that eventually leads to the finished product: furniture, houses, sporting goods, or any of the many applications required by modern society.

In Australia the essential difference is that while forestry is almost entirely carried out by government departments, logging and sawmilling are carried out by companies in the private sector. In Queensland the forestry department did become involved in this activity between 1920 and 1933 when it was responsible for a number of state-owned sawmills. But this was the exception and since then the two activities have been quite separate with forestry being carried out by the public sector and milling, the industrial process, by the private sector.

In spite of this, sawmilling has long been regulated by the forestry department or by other branches of government. Since 1936 all sawmills in Queensland have had to be licenced, for example, so that the requirements of the industry do not exceed the forest resources.

Above all, the forestry department has always controlled the supply of timber from State Forests and because of that it has largely controlled the economics of sawmilling.

The department has had many methods of releasing and controlling the supply of timber and some have been extremely complicated. But basically the department sells by way of licence to a sawmiller timber from a specified area of a forest or plantation. The department grows the trees and manages the area and decides, on the basis of sustained yield in the case of a native forest, how much timber the sawmiller can remove. The department then marks individually those trees that can be felled and the direction in which each has to fall.

In a plantation things are a little simpler. The department plants and manages the plantation and will eventually allow the sawmiller to take selected trees as merchantable thinnings. These are trees which the department wants to remove as part of the silvicultural process and which are now large enough to be milled. Finally, when the plantation has reached maturity, the sawmiller is allowed to clear fell the defined area. After that, and usually very shortly after that, the department prepares the ground and replants it with seedlings

Splitting slabs for a humpy in 1911.

to start the next rotation. There is, inevitably, plenty of potential for conflict between foresters and sawmillers, especially with native forests. For a milling business to be profitable it needs to have a certain volume of timber going through the mill. If the supply is less than this the mill will run at a loss, and if it is greater it should show a profit. But much of that supply is determined by the department, and it determines it not on what is good for the mill but on what is good for the forest.

Again, with plantations things are a little simpler and certainly more predictable. Even as the seedlings are being planted the department knows (provided there are no wildfires, cyclones or other natural disasters) when commercial thinnings will become available and their volume, and when the plantation will be ready for felling and how much it will yield.

Nevertheless, the relationship between foresters and sawmillers has not always been an easy one. Sawmillers often criticised foresters for their aloof and paternalistic attitude, with some justification, while foresters often accused sawmillers of not making the best use of the resource that was available to them, again with some justification. Underlying this, though, was the awareness that each needed the other.

Sawmilling started as a bush industry, and in many respects it still is. Although many mills are now computerised and fully automated, others use more traditional methods and rely on human judgment and muscle. As an industry there is therefore a very wide range of technology, from those with almost none at all to those who use it fully. This wide range is now uncommon in most industries, where new technology has to be adopted in order to remain competitive.

Whatever the level of technology in a sawmill, however, the bush ancestry is never far away. One person might be wearing a leather apron and sizing up a piece of wood by eye, another might be working in an air-conditioned control room and never touch the wood all day. But they are both doing the same thing, albeit in different ways, and what they are doing is as much a part of the bush as forestry itself.

The main function of the sawmiller is to remove trees from the forests and then to saw them into boards or peel them for plywood or veneer. These last two are mechanical processes, as is the manufacture of fibre board, but sawing timber into boards is the activity which retains sawmilling's connection with the bush. This is what sawmillers have always done and, with leather apron or computers, it is what many of them still do.

A sawmill at Chinchilla in 1936.

Although the sawmiller is responsible for removing the trees and then sawing them, in practice the miller concentrates on running the mill and usually employs a contractor to bring the trees from the forest. Most contractors are self-employed and because the work is arduous and dangerous, especially in native forests, many have become legends of the bush.

The first stage in the whole process is cutting down the tree and until relatively recently this was done with hand tools. Cutting down a large, mature tree by hand in a native forest was a skilled and dangerous job. Techniques varied but they had some things in common. First, the cutter cleared a safety area around the tree by cutting away the undergrowth, vines and saplings. This also cleared any snakes from the area. The structure of the tree was then examined, especially in relation to the direction of its fall. If the tree was not uniformly balanced it would have a tendency to fall in one direction rather than another and this had to be taken into consideration.

If the tree had buttresses, the cut would be made above them, where the trunk became cylindrical, which might be three or four metres above the ground. A crude ramp of saplings could be laid

177

over the buttresses to get to the position for the first cut, or steps might be cut into the sloping sides.

The first cut was V shaped and it was made with an axe on the side of the tree facing the direction it would fall. Springboards were then put into notches which were cut on either side of the tree on the opposite side to the first cut. The springboards were positioned so that they brought the two cutters into the correct position for sawing through the trunk.

Standing on the top boards, the cutters then made a series of V cuts with axes to prepare a position for the two-man saw. The saw was then inserted and worked with strong rhythmic strokes towards the original V cut on the other side of the trunk. On a big tree the two cutters were often unable to see each other.

As the saw approached the V cut the timber would 'speak' and the saw cut slowly opened. The sounds from the tree were an indication that it was ready to fall, but it would not go just yet. Usually a few more strokes of the saw were needed and the cutters had to judge precisely when to leave the saw and jump to the ground.

The tree drops slowly to start with and then gathers speed before hitting the ground with a mighty crash. Then there is silence and foliage drops from the newly opened gap in the canopy.

A cutter without a mate was still able to drop a tree like this by using some ingenuity. One method was to fasten the other end of the saw to a vine or sapling and use its natural flexing action to work the other end of the saw.

The modern technique with a chainsaw is not all that different. Two cuts are used to make a V on the side of the trunk which faces the direction of fall. The V is knocked out of the trunk and the saw is taken to the other side and cuts across the trunk a few centimetres higher than the V. When this cut approaches the first the difference in height makes the tree break and it falls in the required direction. Good cutters usually drop the tree within about five degrees of the marked direction.

The next job was to remove the top branches and then to move the log to a ramp from where it could be sent to the mill. This is called snigging and for many years it was done by bullocks. Chains were passed around the log and held by clamps and the bullocks dragged it end first to the ramp, which was often some distance away. Transporting the logs to the mill was more or less difficult depending on how far away it was and what was in between.

At the simplest the logs were loaded onto a vehicle and pulled to the mill by bullocks or horses. But it was rarely that simple. Occasionally logs had to be sent down a watercourse, in which case

SAWMILLING

A charge of timber entering the department's experimental kiln in 1938.

they were left until the next flood sent them down. From Fraser Island the logs were made into rafts and towed across to the mills on the Mary River. And if none of these were possible a portable sawmill was erected in the forest.

Mechanisation changed much of this. Two cutters using a cross cut saw could cut about 8000 super feet a day in a good stand, but with a chainsaw the same two would cut between 20 000 and 40 000 super feet. Tractors were used for snigging and powerful trucks took the logs to the mill.

Mechanisation also made it possible to work parts of the forest that would otherwise have been impossible. Very steep slopes, for example, could make snigging by bullocks or even tractors impossible. This is now done in plantations by a technique called skylining. An aerial line is run from a convenient place on the ridge down to the logging site below. The logs are fastened to the lower end of the line and they are then winched up to the top, rather like a flying fox. On the way they are often many metres above the ground and well clear of standing trees.

The sawmills themselves originally used a combination of steam-driven machinery and muscle but they were a considerable improvement on the earlier method of pit sawing. Once sawmills were available, pit sawing was used only when there was a good stand of trees close to where a building was to be constructed.

In the mill, the logs were first taken to a breaking down bench which had one or two vertical saws. The log was manoeuvered onto trolleys so that it could be passed through the saws, which cut it into flitches. These were slabs of varying cross sections, reflecting that of the trunk and were as long as the original log.

These flitches were also put onto trolleys and taken to the circular saw benches, where they were cut into boards. The flitches were moved through the saws by a series of powered rollers. A gauge was set by the benchman to keep the cut even, and the flitch would be sent back for additional cutting until no more boards could be cut from it.

These early mills used steam to drive the machinery and this was produced from the mill's waste timber. Belts and shafts coupled the machinery to the engine, just as they did in a shearing shed.

There are sawmills today still working much like this, although few now use steam and most have machinery for moving logs and flitches. But the essentials remain the same in that the sawyers at the bench decide how the cuts will be made, and then proceed to make them. The amount of sawn timber produced by a single log is therefore the result of human judgment and it requires a high degree of skill. Indeed, the profitability of the mill is largely determined by these decisions made at the benches. Mechanisation, and especially automation, however, have made a profound difference, particularly to the milling of softwoods.

In the log yard fork-lift trucks are used to move the logs and the work is quicker and safer than when it was done with cant hooks and winches. Saw benches are now automatic and the heavy manual work needed to turn heavy flitches is no longer required. The benchman controls the saw with a series of levers, some of which set the position of the gauges. Indeed, the benchman will probably be some distance from the bench now that the flitches are moved by machinery. Today, the benchman is likely to be found in an air-conditioned control room from which he can see the entire operation.

The use of machine handling also means that the benches can be arranged in a continual line, so that the work is passed automatically from one process to the next. Logs enter the building at one end and emerge from the other as sawn boards without having been touched at any stage during the process.

Much of the work is now controlled by computers. At each stage sensors measure the log and the computer decides the best way of cutting it for optimum yield. The operator can override the computer's decision but rarely has to. Whatever is decided, the saws and the log or flitches are automatically positioned so that they are

SAWMILLING

Snigging a log by tractor at Maroochy in 1939.

cut exactly as determined, and the cut pieces automatically returned and repositioned for further cuts if necessary.

It is an amazing process. Logs are lifted mechanically into a machine to be debarked and the computer produces a picture on the screen showing how it suggests the log should be cut into shorter lengths. If the operator accepts this, the log is positioned and the cuts made. These lengths are then automatically dropped into bins each holding logs of almost identical size.

Another machine takes logs from one of the bins to the start of the sawing line. Again, the computer examines the shape of the log and shows a picture of the log and the proposed cuts to break it down into flitches. When this is accepted the piece is positioned and the cuts made.

The flitches are then automatically passed to the next stage where a similar process saws them into cut boards, which are passed to the final stage where they are cut to length. The boards are then sent to the kilns for drying.

Many pine boards will contain knots, which may be undesirable for certain uses. If these are to be removed, the board will be marked with a special crayon and then passed through another machine. This has sensors which 'see' the marks and cut the board accordingly, removing the small sections that contain knots. The little pieces are called rocking-horse arseholes.

The knot-free pieces that are left are now of random length, but they can still be used. The ends of each piece are cut into a zigzag shape of small fingers and the pieces glued together to make longer lengths. The joins are very strong and almost invisible and these

181

lengths can be used to make structural timbers and moldings such as door frames and cornices which will eventually be painted. There will then be no sign of the join and the result will be just the same as if they had been made of a continuous piece of timber.

In a modern mill there is almost no waste. The bark is removed for landscape gardening, the sawdust and planer shavings are burnt in the kiln furnaces, and practically every piece of timber is put to some use.

Most modern mills have grown out of mills that have been in existence for many years. They are located close to their resource and they have a supply of skilled labour. But the nature of the mill often restricts the amount of technology that can be applied and the result is usually a combination of old techniques and new. The most modern mill in Queensland, however, was built on a green field site at Tuan near Maryborough and opened in the late 1980s. It was designed specifically to use the very latest technology.

The owners, Hyne & Son, are the current generation of one of the oldest and most successful sawmilling families in Queensland. But that success was not easily won.

R. M. Hyne was born in Devon in August 1839, the son of a yeoman farmer. He arrived in Brisbane in 1864 accompanied by his wife Elizabeth. Hyne moved to Gympie in 1868 and towards the end of that year he was advertising in the local paper as a builder and contractor. By 1870 he was making furniture at premises on Commissioners Hill and later that year took over the Mining Exchange Hotel. In 1873 he moved to Maryborough and took up the Royal Hotel. He was mayor of Maryborough from 1878 to 1879, but when his wife died in childbirth that year he left Maryborough and travelled throughout Queensland.

When he returned he had decided on a completely new venture. Aware of the good timber on Fraser Island, he now planned to open a modern sawmill in Maryborough. In 1882 he went to England to buy machinery and when he came back he opened the National Saw and Planing Mills, most of which had been built while he was away. This was situated in Kent Street and had a deep waterfront to the Mary River. The mill itself was the best of its day and the *Maryborough Chronicle* enthused:

> The various machines [built by Robinson & Co. in Rochdale in England] are well worth inspection, commencing with the powerful crane at the end of the mill next to the river, then, following the course of the logs which it lifts into the mill, through a powerful breaking-up saw, then to a travelling bench over 80 feet long, capble of submitting sticks 40 feet long to the action of a 72-inch circular saw. Other saw frames, including perpendicular, cross-cut,

Using a broadaxe to cut railway sleepers in the Brisbane Valley in 1939.

scroll-cutting, are there, also a splendid machine for tongueing, grooving, planing etc . . . and without doubt a machine without its equal in the colony . . . The whole of these machines are driven by a powerful horizontal engine fed by steam from two large boilers set in substantial brickwork.

The mill changed its name to National Sawmill some time later, and if anybody wondered how Hyne could put all this together he readily told them. If he had a family crest, he said, it would be 'an overdraft rampant over a mortgage quiescent'.

In 1888 the firm of Hyne & Son was started when he took his son Henry James into partnership, although the mill still retained its previous name.

Hyne's magnificent mill had been built about two metres above ground level to keep it from the reach of the flooded river, however this was not enough to resist a record flood which came down at the end of January in 1890: 'Mr Hyne's National Sawmill even now presented an appalling appearance, the timber being all crammed in among the machinery, while the river front of the mill was partly burst out from the force with which the current drove the timber against it. One of the corners of the Kent Street frontage carried away and hung over the office, which was carried bodily right into the mill, and was sheltered by the mill's roof.'

At that time there was no insurance for flood damage and the loss had to be taken. Only eighteen days later Hyne & Son announced that they were back in operation. Their situation improved when they won a large contract to supply sleepers for the Normanton railway line.

Unfortunately there was another big flood in 1893. On the first day the buildings were battered but still standing but by the end of the following day they had been swept away and the water poured across empty ground. The damage was estimated at £20 500. This time Hyne had to rebuild a completely new mill, and he did this well above flood level.

He also went into shipping, partly as a separate business and partly to freight timber from the mill. He had a steamer, the *Hopewell*, built in Scotland and it was in service from 1900 until 1934.

R. M. Hyne died suddenly on 5 July 1902 and the firm passed to his son, Henry James. In March the following year Henry James went to America for health reasons and while there he became interested in a band saw mill. The machinery was expensive, however, and he saw little chance of being able to install it at Maryborough. However, in 1905 he was able to buy a complete Canadian plant very cheaply from a mill near Tea Garden in New South Wales which had found it unsuitable for its operations.

This Waterous band mill was probably the first to be used successfully in Queensland to saw native hoop and bunya pine. It remained in service until the end of 1978, although apart from the frame little remained of the original machinery.

Because of ill health Henry James had the firm on the market almost constantly between 1905 and 1920 without a single taker. The industry was depressed for much of this time and the business, like many other mills, was probably struggling to survive. He died in 1936 and his son Lambert took over the running of the company. He had joined the business when he married in 1928.

During the war Lambert realised that the heavy demand for timber was threatening to cut out their traditional area of supply and he began to look elsewhere. In 1942 he bought the licence of a mill in Mundubbera which had recently been destroyed in a flood. He obtained permission to move the licence about 50 kilometres to a site on Brovinia Creek. Within a week he had a boiler and steam engine on site and the mill was operational shortly afterwards. In 1948 he opened another hardwood mill in the Monto district.

He then turned his attention back to Maryborough and pine. Knowing that the native pine would not last much longer he suggested to Grenning that the department put to auction the rights to thinnings from the Amamoor plantations. This amounted to about a million super feet a year with rights for a further nine years. The sale was held in 1947 and Hyne was the only bidder. He built a mill at Amamoor which was specially designed to handle this new, small stock.

SAWMILLING

A breaking-down saw making the first cuts in a log.

In 1949 he bought more forestry quotas at public auction and built a new mill at Imbil on the site of the old state sawmill. It was time for expansion and he continued to buy plantation quotas as they became available and built new mills to process them.

In Maryborough, however, they were having difficulty adjusting to a new supply. The old band saw had been used solely to cut pine logs but when supplies diminished in 1950 they tried to use it to cut hardwood. It proved to be much more difficult than they expected.

Because of the spring in eucalypt logs the usual technique was to use a frame saw to remove two sides of the log simultaneously. They reasoned that because hardwood was denser than pine the band saw should be run slightly faster. But when they tried this the results were dreadful. Then Lambert read in a journal an account of sawing frozen wood in America and discovered that the sawmillers reduced the speed of the saw. They tried it with their eucalypts and it worked.

This technique, however, required the log to be held firmly and turned frequently between cuts. The existing machinery could not do this profitably and so Lambert designed an entirely new carriage for the purpose of cutting hardwood logs. He went to America in 1956 and had his carriage built by Klamath in Oregon, who named it the Hyne carriage. This was installed in 1958 and it served until

1979. By then carriages were being made in Australia and one of these, based on his original design, was bought to replace it.

Lambert Hyne died in 1985 and the business passed to his eldest son Warren and his brothers Richard and Chris. Warren, at his own request, went into the business as soon as he left school instead of going to university and since then had worked his way through every department. Richard had obtained a degree in engineering at the University of Queensland and Chris, 'despite the gloomy predictions of his schoolmaster', obtained a science degree.

It was this generation of Hyne that saw the introduction of computer technology. Initially this was introduced in the old mill at Melawondi and it produced massive problems. These were slowly solved and by the time the new mill at Tuan was designed there was not much they did not know. Perhaps because of this technology, the staff at Tuan are noticeably younger than those at the Maryborough mill. Indeed all the technological developments in sawmilling relate to softwoods and milling hardwood has changed little since the 1960s.

The loss of their supply from Fraser Island was not a direct threat to their business, especially as they received compensation from the government. Their main anxiety was for their staff, and eventually some fifty people lost their jobs. The town of Maryborough was furious about the Fraser Island decision and residents still find it difficult to understand. There is an almost universal view that the decision was entirely political and that they were the victims.

The Tuan mill and the mill at Maryborough are devoted almost entirely to the production of timber for house framing, and their profitability is determined by the state of the housing market. The Hyne brothers study this closely, but their main concern is the continual availability of timber for their mills.

The new mill would certainly amaze R. M. Hyne, with his steam driven horizontal engine. But apart from the use of computers, there is much he would recognise. The shape of the timber being cut from the logs is pretty much the same as when he was doing it but is now done with more precision and considerably faster.

Perhaps that is why the industry still seems to recognise its bush origins. The log goes in at one end and sawn boards come out the other. They always did, even when the mill had no side walls and relied on steam. The changes are with what happens in between, but those changes have been very profound.

John Crooke
Sawmiller

If you met John Crooke in Brisbane you would probably not pick him for a sawmiller. He is a big, outward looking and cheerful man who could well be a marketing executive for a leading company. But this is deceptive. Behind the city suit and urban comfort is a sawmiller who is as tough as the wood he produces.

John Crooke was born in Brisbane in 1935 and was educated at the Church of England Grammar School. When he left he did six months national service in the navy before going into his father's business in 1955.

In 1935 his father had started a hardware business in Brisbane called Queensland Builders Service and ran this successfully until about 1941 when, at the age of forty, he joined the Australian Navy.

During the war one of his earlier customers, the Straker family, saw that there would be a great shortage of timber when the war was over and suggested that he go into partnership with them to take up the timber rights around Allies Creek which were soon to be released.

Crooke senior agreed and the Strakers went to Allies Creek to examine its potential. They found that there was more than enough good timber to meet all their requirements, but the conditions imposed by the forestry department were a problem. Only millers currently holding a sawmill licence could apply and, even more difficult, the timber had to be milled on the site rather than being transported to the nearest town.

The partnership bought a sawmill at Kingaroy, and thus acquired a licence, and then set about building a mill at Allies Creek. They had only hand tools and faced all the shortages of material that were common after the war. Crooke did what he could to supply hardware from his business in Brisbane, which was nine hours drive away over dreadful roads, and Frank Straker built the mill using little more than a pick, shovel and a claw hammer. When it was finished, in 1945, they subcontracted the building of three or four houses. Indeed, they had to build a complete settlement, as there were no existing buildings.

In 1952 the entire mill was destroyed by fire and they had to start again. They went to Cairns and bought the equipment of the old power house which had been built in 1929 and which consisted of steam boilers, steam engines with electric generating equipment and switchboards, and all the associated pumping and preheating equipment. This was moved to Allies Creek where it was used to produce steam-generated electricity, which was very progressive for such a place at that time.

It was this mill that John Crooke joined in 1955. He was no stranger to it, having been there many times on school holidays, and in spite of the lack of comforts he settled in well. 'That wasn't any problem. The toilets weren't flushing, mind you, they were full of sawdust. The showers consisted of a 44-gallon drum under a tank with a pipe of steam going through it, which crackled while you were heating the water. You could shower with as much water as you liked because it all ran back into the dam and the water was covered in dirt anyhow, but you came out a bit cleaner.'

The mill consisted of two benches with steam engines on each spindle and a Canadian driven by electricity produced in the powerhouse. Most of the machinery was Australian made and included a Finlayson carriage from Tasmania and steam feeds from Coffs Harbour. Other items such as guides and rollers were made by engineering works in Brisbane.

The sawing was done by eye at the bench. 'They had to keep the wood straight and on size, and watch out for things thrown out by the saw. It was skilled, hard work. Highly skilled when they had to set the saws up. They had a piece of greasy hemp packing and a cigarette paper to make the saw run warm in a certain spot so it would run straight.'

In 1964 the partnership broke up. The Strakers took two mills which had been acquired earlier in the Eidsvold area, and the Crookes retained the mill at Allies Creek.

John Crooke was now spending half his time at the mill and half in Brisbane. The family still ran Queensland Builders Service until they sold it in 1980 and it marketed much of the wood from Allies Creek. Having sold that company the wood is now distributed to timber wholesalers throughout the state and beyond.

Allies Creek draws its timber from the Boondooma State Forest, which is a native forest producing spotted gum. It is, says John Crooke, some of the finest spotted gum grown anywhere. 'Our allocation area is producing wood that is better than average shape and it is beautifully clean and free of defect. It should be going into furniture, and it will one of these days.'

The mill draws from a very big area but there is not much standing timber in each hectare. Logs might have to be brought nearly 30 kilometres from one direction and up to 180 kilometres in another.

There are about sixty people living in the settlement, of which about twenty-four are employees and the rest are contractors, including two cutters, a snigger hauler loader and a couple of truck drivers.

The settlement is still isolated and self-contained and many of those working there now were born there. It is also more comfortable than it was. 'It was a pretty wild sort of place in the 1950s and the police knew the road to Allies Creek very well. But people have mellowed a bit now, built fences around the place and so on. They enjoy the life. It is certainly different.'

John Crooke introduced computers to the mill in the early 1980s and found that the younger staff took to them with complete assurance. It is still not as automatic as some mills, but the use of computers has made a big difference. 'It has given us a straighter, cleaner and more accurately sawn product than we had before, with less defect in it. The old breast bench man with his thinking cap on and a leather apron would still cut you a very good piece of wood. It would be superior to what we are cutting now, except that we can cut it ten times faster and with no fatigue.'

As for the future, he sees that much of the framing market will be supplied by exotic pine from plantations and the hardwood market will increase in the use of good quality wood for furniture and as treated wood for landscaping in gardens and recreation areas. 'The hardwood mill in ten years or so will be involved in drying and dressing high-quality engineered products. It is doing this to a large extent now, but by then it will be dominant.'

He is optimistic about the future, especially as much American timber has been locked up because of concern over the spotted owl. This, he says, will give the Australian industry a breathing space to prepare for the future, whether it is in building big new pine mills or modifying existing hardwood mills.

In his spare time John Crooke goes fishing and boating. But not, perhaps, your average boating. He owns a 58-foot motor cruiser which was built in Tasmania out of 50 tonnes of huon pine. It contains two Perkins V8s and has as much refrigeration space as a small hotel. 'It is a lovely piece of machinery that we enjoy from time to time.'

He also maintains his enthusiasm for rugby and is still a keen follower of the game. But, like the people at Allies Creek, he has probably mellowed with time. Which is just as well.

The sight of John Crooke bearing down on you on a rugby field would have left a memorable impression. If you still had a memory.

13 Changing Uses

Because wood is one of the oldest resource materials known it might be thought that its uses are equally long established and well known.

In some respects they are. Anybody crafting wood with hand tools is simply repeating a process that has been performed for centuries. Even the tools are not much different. They might be sharper, and they will certainly stay sharper longer, but most would still be recognised by a medieval craftsman.

Beyond that, however, the use of wood has changed considerably and continues to do so. In some cases it has had to give ground to the use of plastics. Recreational boats, for example, are now nearly all made of fibreglass although its superiority is not quite as great as originally thought. Fibreglass is certainly resistant to marine borers and does not rot, two very big advantages over wood, but it has also been found to suffer from osmosis and this can be very expensive to treat.

One of the newer uses of wood, though, is in the form of fibre board, and especially the newer medium density fibre boards. These provide a relatively cheap and easily worked material for use in building and furniture and allow timber to be used in ways that would otherwise be too expensive or impracticable. Heavy duty laminated timber is now used for weight-carrying beams and these also have a decorative effect which other materials lack.

Wood varies considerably from one species to another. Some are

Rotary lathes producing sheets to be made into plywood.

easy to work, others take stain or glue well, others have good mechanical properties such as strength and will not collapse when under load. Others might have none of these qualities, or combine some while lacking others.

It is obviously important that suitable timber is selected for the purpose for which it is being used and that it is available in sufficient quantity. Most people who regularly use wood are aware of the qualities they need and the species that provide them. But in many cases there are other woods, less well known, that will do the job even better than those they regularly use.

Examining the qualities of different species of timber is the subject of continuous scientific research throughout the world. However, much of this research is regional in nature as it focuses on timber that is available in that area. So while the total research knowledge is considerable, little is of universal use and much of it is very specific indeed.

In Queensland a great deal of research has been carried out by the forestry department into the properties and use of Queensland timbers. This research is a continuing activity because of the changing information that is required by the industry. If a manufacturer develops a new glue, for example, the industry needs to know which timbers it can be used with and which cannot, otherwise there could be some very expensive mistakes.

This research is undertaken by the forestry department, either to extend its own knowledge or for outside clients who pay a fee for the work to be done. This research has nothing to do with growing trees—it is about getting the best out of them after they have been logged.

Once again it was Swain who led the way. In 1920 he opened a Forests Products showroom in Brisbane where users of wood could discuss their problems with trained staff and see samples of Queensland timbers. It was a 'shop-front' service and was very popular even with curious members of the public.

The following year he established the Forests Products Bureau and this concentrated on research into wood technology. This activity is still carried out today although there is no longer a Forest Products showroom. One of the main aims of the Bureau, and particularly its wood technology branch, was to encourage the use of lesser known species of timber.

This has been a problem for many years. Users became familiar with the prime species and wanted to use them all the time. This in turn put a heavy strain on those species in native forests, while other species which might be equally suitable were never called for. The prime species were usually difficult to regenerate, and certainly in the quantities demanded by the market, and so they were always in danger of being over-cut.

Plantations were developed to provide the market with pine, but there was still a heavy demand while these plantations were reaching maturity. The hardwoods, which did not lend themselves to cultivation in plantations, were under threat all the time.

Queensland is so rich in timber species that there is nearly always more than one species capable of doing the job equally well as that traditionally used. Swain's concern was to make that widely known and to encourage their use and thus spread the load on native forests and lead to their better utilisation.

In 1923 Swain produced his Universal Wood Index which classified all the commercial timbers produced in the state and compared their characteristics with exotic timbers. When the Index was finally revised in 1926 it listed 250 species of Queensland timber. This was supported by the publication of Swain's *The Timbers and Forest Products of Queensland*.

Although Swain started this programme of research and public education, it continued and expanded after his dismissal. Wood technology now included biology, seasoning, treatments for preservation and many other subjects.

As for the industry, some changes came about through need while others were a result of specific research.

One that came about through need was the use of maple as a substitute for cedar. By the time cedar had become rare as a result of reckless cutting, maple was found to be almost as good. If this had been recognised earlier it might have reduced the cutting of cedar and so kept it available for much longer.

An example of change as a result of research was the use of walnut. Although this tree grew to a great size it was rarely milled because it had an abrasive property that severely damaged saws. Because of this, walnut was considered to be an unusable species and was burnt when land was being cleared.

Research showed that the wood had a significant content of silica, and it was this that was doing the damage. The answer seemed to be to use more resistant saws but attempts to develop these were unsuccessful. What was discovered, though, was that walnut could be cut to make plywood. The cutting was hard on the knives, but there were fewer problems than when cutting into boards.

The plywood did not prove popular, however, and it was only as a result of intense marketing by the department that a market was found in America in 1927. This produced a very strong demand for walnut plywood and even for whole stumps. Unfortunately this trade was badly affected two years later when American buyers complained that they had been sent stumps that had been damaged by borers. The department responded by offering to issue certificates of quality to those exporting walnut and this succeeded in cleaning up the market. Today, Queensland walnut is listed in the highest category of Queensland timbers.

Another essential area of research, especially in promoting the use of lesser known timbers, was the seasoning of timber. One of the reasons why some species were not popular was because little was known about their preparation and qualities and users were reluctant to experiment when a better known species had all the required qualities.

Seasoning, or drying, is crucial to the use of wood and its success largely determines the wood's suitability and length of service. Developing the best methods for the lesser known species was not likely to be carried out by sawmillers or others in the industry, and they were not likely to use them until somebody had.

Swain recognised this as a legitimate function of the forestry

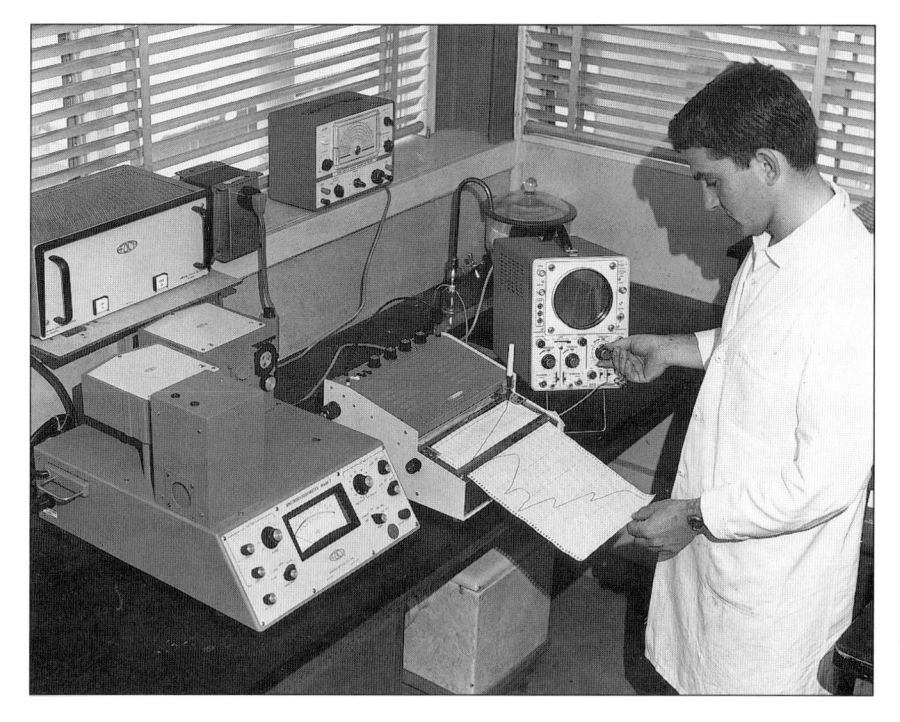

A chemist analysing timber samples at the Forest Products Research Laboratory in 1970.

department. Indeed, it was essential in the context of his programme to bring these lesser known species into wider use. So in 1928 the department set up a seasoning plant in Brisbane to carry out research in this important area and it has been in use ever since.

Trees that are growing contain a significant amount of water, although some have more than others. Hardwoods generally have less than softwoods. The walls of the wood cells are actually saturated with water, and the space inside the cells also contains a great deal. When the tree is felled this water starts to dry out, but not at a uniform rate. The water in the cell cavities starts to dry out first and it is only when the cavities are dry that the water in the cell walls starts to evaporate.

Water dries much more rapidly from the ends of timber than it does from the sides. When the water dries from the cell walls they start to shrink, and as they do so they set up stresses inside the wood because the cell walls in the inside of the wood are not yet dry and are still in their original form.

These stresses can produce cracks in the surface of the wood and even inside if the forces are strong enough. The quicker the wood dries, the greater the risk of the log or board being degraded in this way.

The wood is said to be seasoned when its moisture content is in balance with the normal humidity of the atmosphere. When it reaches this state any further movement will be a result of atmospheric conditions and the wood is therefore stable. A few other changes will have taken place as well. Most species are considerably stronger and lighter when they are seasoned. They will also have shrunk slightly from their original size. If wood is used for joinery or building before it is seasoned, gaps are likely to appear at the joins as the wood continues to dry.

Some Queensland timbers are not seasoned but are supplied 'green off saw'. This is because they become so hard when seasoned that they are almost impossible to work.

The equilibrium that exists when the wood has been seasoned is maintained only if the atmosphere around it does not change, or at least not to any great extent. This is the case if the wood is used in a similar climate to where it was seasoned, but it is no longer the case if wood seasoned in the tropics is used in temperate climates or if the atmosphere is very dry because of air conditioning. Seasoning can be adjusted for these events provided they are recognised and provided the wood is maintained at the appropriate moisture content.

The traditional method of seasoning wood was to put it outside, usually with a temporary roof over it to keep off the rain. There is nothing wrong with this method and for a long time people preferred it even when there were newer alternatives. Air drying suffers from two weaknesses. One is that it takes a long time. A board 25 millimetres thick can take between three and nine months to dry naturally depending on the species and the weather. The other weakness is that the wood can be seasoned only for the local atmosphere and further drying will be necessary if it is to be used in a drier atmosphere.

The alternative to air drying is to use heated kilns. These are built of brick or concrete and are essentially large steam-heated ovens. A stack of wood is placed in the kiln and the steam drawn in by a fan. The stack, and the kiln itself, is designed to give an even distribution of the heat, so that each piece of timber receives exactly the same treatment.

The use of a kiln offers complete control of the seasoning process and the local atmosphere plays no part. Both temperature and humidity can be controlled but doing so is a skilled job because the drying conditions have to be uniform throughout the kiln. This means that the wood can be dried to almost any moisture content required by the end user.

The moisture content is monitored all the time the wood is in the kiln. This is done by thermometers which measure heat and humidity or by electronic probes which measure the moisture content directly. Sample boards are also used. These are short pieces of the same timber and their ends are coated so that they will not dry quicker than the rest of the stack. These boards are removed from time to time and their moisture content determined by weighing.

Kiln drying is much quicker than air drying. If 25 millimetre thick hardwood is given a short period of air drying first it can be seasoned in a kiln in two to four days. Many softwoods can be put into the kiln as soon as they have been sawn, with obvious economic benefits.

The forestry department has done much to encourage the use of kiln drying in the industry. Its facilities at Salisbury in Brisbane have been widely used to train industrial operators as well as conducting many experiments to determine the best method for different species. Indeed, this is a good example of how knowledge has to be obtained locally to supplement research that has been carried out elsewhere. The requirements of Queensland species are likely to be determined only in Queensland.

This branch has recently been engaged in experiments with high temperature drying of house framing timber from plantation pine. When the first experimental work was done with kilns at Salisbury about twenty years ago it was found that a temperature above boiling point (120 degrees Celsius) kept distortion to a minimum. The techniques developed at that time were adopted by plantation industries throughout the world. At this temperature the timber dries in about twenty hours.

Later developments in the design of kilns and extensive knowledge about their use have led to experiments with temperatures as high as 200 degrees Celsius, which produces a drying time of about three hours. These experiments were commissioned by Hyne & Son and the kilns at their new mill at Tuan are now operated at temperatures close to this with a considerable improvement in efficiency.

While developments in seasoning might be an example of better use rather than a different use, much of the Queensland timber industry is based on a use that was not known to the medieval craftsman: plywood and veneer.

Plywood and veneer are produced in sheets of varying sizes, they are strong for their weight and in the case of veneers they can offer a surface of high quality cabinet wood that would be prohibitively expensive if it were solid.

Plywood mills were developed in Queensland during the First World War and they used hoop and bunya pine, both of which were

A press advertisement run by the Forest Department from 1979 to 1988 to locate infestations of the West Indian drywood termite.

then readily available. After the war other timbers were brought into use, including walnut, maple and satinay. In producing plywood the log is 'peeled', that is, it is rotated against knife blades so that the wood comes off in a continuous sheet of constant thickness.

In 1930 Cairns Timber Limited established the veneer industry in North Queensland when it installed a vertical veneer slicing machine. This sliced thin pieces off the log and its many applications led to the wide use of rainforest cabinet woods. Most of these veneers were exported.

By the 1930s the Queensland timber industry was producing more than 80 per cent of the entire Australian production of plywood and veneers. Their production is still a very important part of the Queensland timber industry although today only plantation timber is used.

The latest development is structural plywood, which is considerably thicker than the familiar variety and which is now used for bracing walls in house construction. Laminated beams are another interesting development which has helped to maintain the amount of timber used in building after concrete slabs reduced the need for hardwood flooring.

The manufacture of particle board has also done much to make the best use of forest products. Particle board originated in Germany after the Second World War. It is made by mixing small particles of wood with a binder and then compressing them under heat. The method was born out of a need to use sawmill waste and at a time when large sawn boards were almost unobtainable in Europe.

The first particle board factory in Queensland was opened by Jim Wayper, a Gympie sawmiller, in Brisbane in 1971. When plantations came into production he opened a state-of-the-art factory at Gympie in 1975. The company was bought by CSR in 1981 and it is now a leading manufacturer of particle board in Australia.

Each board is made up of three layers: two surface layers and an inner core. After much trial and error these are now made from flakes of timber which are sorted and treated for use in the different layers. The success of particle board was based on its uniformity, which was rare in a product made from something as varied as timber.

A more recent development is medium density fibre board, which is also made at Gympie. The log is converted into pulp and this is fluffed up, coated with glue and then compressed. This produces a board which is very stable and which can be worked and well finished. It can be edge-moulded to a fine finish and used in cabinet and furniture making. It can also be used as a base for laminates and veneers. Although the high manufacturing costs make it more expensive than particle board its value to forestry is that it can be produced from plantation thinnings, of which there is now a considerable supply.

Because foresters habitually think in long timescales they are often in a position to see trends well before the timber industry becomes aware of them. They can therefore alert the industry and help them prepare for the future. An example of this was the large amount of softwood that became available from maturing plantations combined with the effect of the ban on logging North Queensland rainforests as a result of World Heritage listing, and the hardwoods that were still available from forests in the south-east of the state.

The department mounted a campaign to persuade the industry and manufacturers to establish furniture making in that part of the state and to use these hardwoods as an alternative to rainforest timbers that were no longer available. Most hardwoods had been

GROWING UP

Harris Court in George Street ready to be fumigated to eradicate West Indian drywood termites.

traditionally used for house framing but this could now be met by pine plantations, thus freeing the hardwoods to be used in more highly valued products. In encouraging this change, the department was continuing the direction that had been set by Swain: to use a wide range of timber species and to use them in a manner compatible with their quality.

It can be seen from this that many activities of the forestry department are far removed from growing trees. The department is closely involved with all uses of wood and the general well-being of the timber industry. It is perhaps one of the reasons why the relationship between foresters and the industry is probably better now than it ever has been. Although not always agreeing with each other, foresters are now more aware of the needs of the industry, and the industry recognises that the research carried out by foresters has done much to improve their efficiency and profitability.

Wide ranging though the department's activities are, some of its activities seem to be carried out simply because they are the only people with the necessary skills even though they are beyond the traditional skills of forestry. An example of this was the department's investigation into the arrival of the West Indian drywood termite, *Cryptotermes brevis*.

This is a very destructive pest which comes from the Caribbean and which can literally destroy timber in buildings. Its voracious attacks had been well recorded in the Caribbean, in South Africa and in parts of the tropics but it had not been a threat in Australia. This was fortunate because although there are plenty of Australian termites to wreak havoc on timber none are quite as destructive as *C. brevis*.

In 1966 *C. brevis* was discovered at Maryborough and investigation revealed that it was well established and had been there for many years. Some ten years later it was also found in Brisbane.

The termite usually attacks furniture and timbers in sheltered parts of the building such as roof timbers, internal walls and floors. It forms large colonies which can remain undetected for years, as they had in Maryborough. Unlike other termites, *C. brevis* does not need contact with the ground.

It was assumed that the pest had arrived in Australia during the Second World War when a considerable amount of American military supplies passed through Brisbane docks. It had then been spread in furniture and other items that had been sold when the war ended.

Because the presence and elimination of this unwanted visitor did not fall into the activity of any other government department, the forestry department was asked to investigate the problem and, if possible, get rid of it. The pest was eating timber, after all, and forestry specialists knew a great deal about timber pests.

The first task was to find out how widespread it was and in 1974 the media were used to alert the public to its presence. This raised considerable interest and it was followed by an intensive advertising campaign which ran at certain times of year from 1979 to 1988.

The best way of identifying the presence of *C. brevis* was by its droppings, called frass. Householders were asked to send in any droppings that looked similar so that they could be examined by the department's scientists. The response was almost overwhelming and many houses were identified as being infested.

The treatment used was fumigation. Those living in an infested house had to leave, taking as few possessions as possible. A contractor then sealed all the openings and crevices before pumping in the fumigating gas. The gas was kept to a required level for a whole day and then fans were used to disperse it. When the fumigant had been cleared the premises were certified and the inhabitants were allowed to return. The whole process usually took five days. People were paid compensation and costs but they could not decide against fumigation. Once the pest was discovered fumigation was obligatory.

When some of Brisbane's largest public buildings were found to be infested, they too had to be fumigated. Working on buildings this size produced all manner of problems, but at least those living and working in the city had the unusual sight of some of the bigger buildings being completely wrapped in plastic.

The whole of Parliament House was fumigated in an operation that took nineteen days to complete. During the work the contractor used 16 kilometres of sheeting 2 metres wide which was made up into large sheets, 15 000 clips, 6000 G clamps, 50 tonnes of sand, 7 kilometres of gas tubing and 2.2 tonnes of fumigant.

Cranes and cherry-pickers were used to put the plastic sheeting on the roof and workmen abseiled down the walls to unroll the sheets and clip them together. Fans were used to remove air from inside the building so that the plastic was drawn tightly against the outside of the building. Other buildings in central Brisbane that were treated included the Government Printing Office and the old Queensland Museum. Since 1978 a total of 161 buildings have been treated, including thirty-three government buildings. The treatment has proved to be very effective and the number of infested buildings fell considerably after 1988. Those that were found were isolated cases, in contrast to the earlier pattern where many nearby buildings were also infested.

By 1988 this work had cost the State Government $1.6 million but this seems cheap against the US$105 million estimated to have been spent in repairing and fumigating infested buildings in Florida in 1983 alone. There are still occasional infestations in Queensland and the department is still responsible for the treatment.

Meanwhile the research branch of the department continues to investigate almost every aspect of timber use. It conducts extensive programmes on kiln use, preservation of all timbers, durability trials, timber use surveys and biological research, as well as undertaking specific projects requested by industry.

The results of the department's own research are made available to the industry and there is no doubt that it has been of great benefit to them, to all users of timber, and to the community as a whole.

Andy McNaught
Scientist

Andy McNaught was born in Melbourne in 1955 and took up forestry almost as an afterthought.

When he left school in Blackburn in 1972 he learnt that ANU in Canberra had introduced a scheme by which they would take students on the basis of their entire secondary school performance rather than on their exam results in year 12. After a disappointing final year this seemed to be the only way he could gain a place at university.

His preferences were architecture and engineering, but as ANU did not offer courses in these he selected his next preference, which was forestry.

He spent five years doing the course and found it enjoyable and satisfying. But he was conscious that it had, at that time, little contact with the real world of forestry. He thought the students from Queensland had a big advantage because most had done a year or two in the forestry department before going to university. By the time he finished the course in 1978 the university was producing forty or fifty new foresters each year and of these only about fifteen found jobs in forestry departments. Fortunately the Woods and Forest Department in South Australia at that time had an enlightened scheme whereby they employed forestry graduates for a year 'to make them employable'.

Andy spent most of his year at Mount Gambier and this was his first contact with real forestry. He spent about six months on harvesting research, trying to improve the first mechanical harvesting systems in Australia. During that time he met the owner of a plywood mill in Victoria who was doing peeling trials at Mount Gambier. Andy was invited to see his main mill when he had the chance. He did so and was offered his first real job.

In 1978 he joined Savaco as logging manager and projects officer at its mill in Broadford in Victoria. His job was to coordinate the log supplies and make sure they were up to specification, which needed good contacts with other companies and the Victorian department.

He also became involved in examining new projects: finding new processes, costing them, calculating labour requirements and finally presenting recommendations to the board. In 1980 he became production manager for the plywood mill but the following year he decided to leave Savaco and return to Canberra. There he returned to the forestry department at ANU as senior technical officer working in timber testing, drying and preservation.

He stayed there for three years but in the end he realised that he was not extended as much as he needed. This suddenly changed when he was asked to join a new company called Auswood as a consultant. This had been started by a previous managing director of Savaco and offered specialist advice to the plywood industry as well as importing plywood and slicing equipment. 'That was fantastic. Exactly the opposite of what I had at ANU. ANU was staid and the work load was pretty light. At Auswood everything was absolutely full on, to extremes.'

For the next two years it was non-stop action. He studied new machinery from Finland, he advised a drawer manufacturer in Sydney, he advised a Tasmanian mill who wanted to install slicing equipment and did the same for a mill in South Australia and one in the south island of New Zealand. He became particularly involved in developing laminated veneer lumber which could be produced in long lengths for use as supports in buildings.

Exciting though the job was, he now had a young family and was anxious to reduce his frequent absences. So in 1986 he left Auswood and joined the Queensland forestry department as a timber technologist specialising in wood drying. In 1989 he was appointed officer in charge of processing, seasoning and mechanics, and in 1991 he became the Principal Utilisation Officer.

He spends about 40 per cent of his time on research. Twenty per cent is spent helping the plantation production programme and the remainder is spent with outside clients of the department and fighting the inevitable paper war.

His speciality is heat drying of pine. The original research was carried out in the 1960s in America, where it excited little interest. It was noticed in Australia, however, and the work was repeated by the Queensland department at the kilns in Salisbury. Andy McNaught continued this work and investigated the effects of drying at 200 degrees Celsius.

'The real concern from the marketing point of view was what it did to the strength of the wood. So it was not just a drying experiment, it was timber mechanics as well. Structurally it has no significant effects on the strength of the wood.'

While this is undoubtedly one of the success stories, most research work is more mundane. Recent work included a series of trials to investigate the durability of various timbers when they are above the ground in exposed situations, and carrying out sawing tests on hybrid pines which are to be planted in south-east Queensland.

'It is important that sawmillers are not disadvantaged by changes like this. Industry is rightly nervous of change. They have built their mills and have invested significant capital. They don't want the ground rules changed in terms of the resource if it means they have to spend more money. So we carry out these studies so they know what to expect. We are seen by the industry as independent researchers, not simply doing PR work for the department. So when we say that this study has found that there are not going to be any problems they are pretty comfortable.'

In his spare time Andy McNaught concentrates on outdoor activities. 'My main interest is bike riding. I used to do it competitively but not any more. I still manage to crank out 200 kilometres a week on the bike and I ride to and from work each day. It is quicker than the car!'

Andy McNaught thinks deeply about modern forestry. He does not describe himself as a conservationist, but his thinking has much in common with theirs.

'As community expectations and population pressures grow it may mean that you have to pull areas out of production, particularly along the coastal strip. Pull them out for preservation or for other use, or limit the way you manage them. Just because we can produce logs in perpetuity does not give us the right to keep managing those forests for timber in perpetuity.'

Above all, he loves wood. 'A piece of framing is a commodity item, but it upsets me when I see a piece of good timber being used for structural purposes. One of my aims is to see the attributes of really beautiful timbers recognised. A lot of old house frames in

Tasmania were made of huon pine. Now, you could not buy that if you owned a bank.'

Andy McNaught's combination of industrial experience and science training is unusual, or certainly was in the past, but the benefits are very real.

14 Technology and the Forest

Growing trees is not a simple matter. If they are the right trees in the right place they will certainly grow, just as they always have. But forestry is primarily about growing trees for specific and clearly defined purposes which in turn are based on the needs of the community it serves.

In order to do this forestry has to concentrate on two basic themes: it has to produce timber of the required type and quantity as economically as possible, and it has to maintain continuity of supply while the demand exists. In other words, it has to produce trees efficiently for as long as they are needed.

If it can be simply stated, it is certainly not easy to do.

Some of the complexities are common to many enterprises: forecasting market changes, economic pressures, population growth and so on. Others are not so common.

Trees are formed by natural growth which in turn is determined largely by their environment. While foresters can to some extent control that environment they cannot control the whole process. They also have to contend with timescales that are very lengthy and this makes it difficult to respond to a quickly changing market. Every sawlog which will be used in Queensland during the next twenty-five years is already in the ground.

Forestry has therefore come to rely very heavily on technology in order to fulfil its role. Technology is now an essential part of

Measuring a sample tree for yield calculations, Imbil, 1963.

forestry and it is applied in many ways and across many disciplines: tree breeding, the destructive habits of insects, soil chemistry, climatology—all, with many others, are part of modern forestry.

The effect of this technology is not obvious to the casual observer. Native forests and plantations look pretty much as they always did. But that is deceptive. What has changed is that the productivity of the forests has improved considerably. The amount and quality of timber produced by a given area is now much greater than it was, especially with plantations. To a forester, the forests do not look the same as they always have. They look better. Indeed, they are better.

In this chapter we will look at two examples of forestry work which have been revolutionised by modern technology: forest modelling and cartography. Both are as old as forestry itself and both have been largely transformed in the past few years.

Forest modelling is a technique used to predict the growth of a forest. These models can be simple (and before the use of computers they always were) or extremely sophisticated. They are used to provide information on which many management decisions are based.

Predicting the growth of something as complicated as a forest is by no means easy. A forest is a biological entity which does not follow mathematical rules and because of this the application of mathematics, no matter how sophisticated, can never produce a precise forecast. The results of forecasts have to be tempered with a realisation that a living forest might not develop exactly as predicted.

Forest models draw heavily on another technique known as mensuration. Mensuration means measuring and embraces the mathematical rules of determining volume, area and length. But in forestry it has a more specialised meaning. It is the study of the measurement of individual trees, of stands of trees and of entire forests.

With individual trees it includes their detailed measurements and the production of functions to describe the shape of the tree and other important aspects. With stands it includes the actual volume and size assortments of the stand. This can be done on a bigger scale to measure whole forests, the forests of a complete region or of the entire state.

When measuring an individual tree the first thing to be decided is whether it will be measured destructively or non-destructively. In other words, will it be measured while it is standing in the ground or will it be cut down first. Measuring a tree is more difficult when it is in the ground, but the advantage is that it can be measured again a few years later. Destructive measurement, however, is usually much more accurate.

Measuring an individual tree will produce only limited information. So the individual tree is usually part of a sample of trees, each of which will be measured and the results used to provide specific information. The first step, then, is to decide the tree population that is to be investigated and then to select a representative sample of that population. The next step is to decide what aspect of that population is to be studied. This might, for example, be volume in relation to tree diameter. In this case the sample will need to include trees that cover the entire size range.

The height of the selected trees will then be measured, together with the diameter at regular intervals along the length of the trunk. The volume of the tree can then be calculated from this information and the shape of the tree can also be defined.

These measurements are recorded on standard forms and are entered into a computer data base called the Sample Tree Library. This data base includes much more than the measurements just described. Species, age, the quality of the site and the treatment it has received are also recorded, so that the individual tree is related to much broader circumstances.

Provided all these circumstances have been recorded, and provided the sample is big enough and truly representative, then the information can be used in a model to predict the growth of similar trees in similar circumstances. It is this ability to predict that makes modelling so important in modern forestry.

The policy of sustained yield depends entirely on these predictions. This policy requires that the cut from a native forest is limited by what the forest will grow in a similar period, but it will only be as good as the prediction of that growth. If this prediction is inaccurate, then the cut will either be too much or less than it could be.

Sustained yield management is not applied to plantations because there is no intention of keeping a plantation forever. On the contrary, the trees in a plantation will be harvested when they are ready and a new plantation will be established to replace them. There is, however, a need for growth predictions and other information in the management and design of plantations.

In Queensland the sample library of plantation trees contains measurements of about 5000 hoop pine and the same number of slash and Caribbean pine. More measurements are made every year, adding details of several hundred trees to the library.

In practice, measuring diameters at intervals along the trunk of a standing tree is difficult and prone to error. The technique has therefore been simplified so that only one measurement is needed to predict volume against height or age. This measurement is the diameter at breast height and it is calculated from the circumference of the tree at 1.3 metres above the ground.

The accuracy of these predictions depends almost entirely on the size of the sample and the accuracy of the measurements. The sample size for plantations now runs into many thousands and predictions can be made with a high degree of mathematical accuracy, which is itself defined. But, as explained earlier, trees are living beings and their actual growth might not correspond exactly with the prediction.

The samples of hardwood and rainforest trees are smaller and narrower in range. Trees in native forests and rainforests are much more varied than those in plantations both in terms of species and height. Predictions of native forests are always accompanied by an indication of the confidence limit of the prediction. Those making decisions based on the prediction can therefore see how heavily they can rely on it.

The computer data base can be used for many kinds of predictions but most involve volume in relation to other aspects. Volume

Traditional surveying techniques being used in a State Forest in 1963.

in relation to size is one common example, but information may be needed on how that volume is distributed within the tree. This is important in predicting the size of logs which will come from the tree, how they will vary along the length of the trunk and the point towards the end of the trunk where the log will be too small to mill.

Another measurement common in plantations is the site index. This provides a measure of the productivity of the site in terms of its capacity to grow wood. As it is closely related to volume, height is used because it is easier to measure.

The height of a tree obviously varies with its age, but not at a uniform rate. At first the height does not increase very much, then it increases rapidly in the young tree before decreasing as the tree reaches maturity. The site index is determined by the height of trees in the stand when they are twenty-five years old, which is the age when the site will be fully expressed but not too close to maturity when growth will be slowing down.

The height of a stand of any age can be compared against a known standard and is calculated by measuring the height of the fifty tallest trees per hectare. Knowing their age and the standard height expected, the quality of the site can be determined. A good site will produce trees larger than the standard reference and in a poor site they will be smaller. The site index is therefore a measure of the ground itself and its ability to support the trees growing on it.

Another common measurement is known as basal area. In simple terms basal area is the total area of trees contained in a stand of, say, one hectare. This total area will vary from one stand to another and is also an indication of the worth of the site because it can be used to calculate the volume of the stand.

Basal area can be calculated by measuring the circumference of each tree at breast height, using that to calculate the cross-sectional area and then obtaining a total of all the trees in the stand to give the basal area. While this gives a very accurate measurement it is not always practical or necessary to measure every tree. Instead a sample of trees are selected for measurement.

So how does the observer select the trees to be counted? They are selected by their width as seen from the observation position. When this is being done accurately the observer looks through a prism. Some trees will be optically deflected by the prism and some will not. Those that are deflected are the required width and are counted. The others are ignored.

This is based on the fact that trees of a certain width will all subtend the same angle at the eye of the observer in that location, and the prism is used to determine which trees subtend that angle and which do not.

A simpler method, not common today, was often used in the field. The observer extended one arm and raised the thumb vertically. The thumb was then swept across the stand and the trees that were the same width as the thumb were subtending the correct angle and were therefore counted. Although useful for estimates, this technique was never used to collect accurate data.

In both methods, the prism and the calibrated thumb, it is not the actual width that selects the tree—it is the observed width. Thus a spindly tree close to the observer will subtend the angle and will be counted, while a distant giant might not and will be ignored.

Thinning will obviously reduce the basal area of a plot because some trees are being removed. The accuracy of the thinning can therefore be determined by measuring the resulting basal area and a calibrated thumb is excellent for measuring the progress of the thinning while it is being carried out.

Accurate measurements of basal area are also incorporated in the data base. Some will have been made from sample plots in which

TECHNOLOGY AND THE FOREST

Two forestry technicians prepare photographic equipment for aerial photography of a State Forest.

every tree has been measured, while others will have been made in bigger areas with a prism.

When plantations were first established there was very little knowledge on which to base predictions of future growth. Sample plots of a tenth of a hectare were established as part of the plantation and these were used for accurate measurements of all kinds. The measurements also indicated the success or otherwise of different treatments such as the use of fertiliser and weed control.

As the amount of this information grew it could be used in computer models to make predictions about plantation growth. It can now be used to predict the entire future of a plantation even before it is planted, and allowing for soil conditions, climate and many other variables. The data base is searched to match all the known variables and the computer will predict the volume of timber the plantation will yield at maturity, how long that will take, and the volume of merchantable thinnings that will be available before then, and when. All before a single tree is planted.

Given accurate data, the computer model can actually 'grow' the plantation under various regimes of treatment. The resulting yields produced by the model give a comparison of these regimes and for the first time the forester is free from the long process of experimenting in the field. The model will show the best treatment for a species of tree in a specific location over a known period.

The model can do this for a stand, a forest or the entire state. Indeed, the bigger the subject the more accurate the results will be. On the other hand, the further ahead the prediction the less reliable it will be. If the whole of the state's forests are forecast over the next ten years some plantations will have been harvested during that time and replaced by younger trees. The native forests will also have changed during that time. All these changes are reflected in the computer models.

There are two types of models. The deterministic method is the simpler of the two in that it takes the original data and simply predicts it into the future by linear extension. In other words, it is based on the assumption that what has happened in the past will continue to happen in the future.

While this has the advantage of simplicity, it does not relate well to the real world. Over the next ten years, for example, there might be events that have not occurred in the recent past. These include cyclones, wildfire, insect attacks, and all the other things that can seriously affect forest growth. While some of these will be reflected in the deterministic model, because to a varying extent they will have influenced the data collected from the field, their occurrence over the next ten years cannot be predicted.

The second type of model is called stochastic. Stochastic means random, and this method allows the inclusion of many more variables in the model. Measurements of a sample of trees vary widely. In any sample the height of individual trees, for example, will be spread over a wide range from the smallest to the tallest. A deterministic model relies on calculated averages of these measurements, while a stochastic model can include the full range of variation. Because of their complexity, stochastic models take far longer to run and are therefore not commonly used. They do, however, promise more sophistication for the future.

It can be seen that computers have been responsible for massive improvements in modelling and mensuration techniques in the past few years and they have made the process infinitely more complex and accurate than it once was.

Before the use of computers calculations were done by hand or with the use of simple aids such as tables, slide rules and adding machines. These calculations were based on field measurements but all the data had to be manually retrieved and then manipulated in what was a slow and laborious process. There was virtually no way of introducing variables and the greater the sample, the more laborious the calculation.

The first improvement was the use of punched cards which could be read and sorted mechanically. This was followed by the first use of computers in the late 1950s. At first the punched cards were sent to the IBM computer centre in Sydney where they were analysed and the results received the following day.

The next stage was the use of the new computer at the University of Queensland, which was used by a number of government departments. This was followed by the use of a computer which had been installed at the State Treasury and this in turn was replaced by the State Government Computer Centre, which was a very large installation offering much improved facilities. In time, of course, the forestry department installed its own computers.

Each stage saw a vast increase in sophistication, but the use of in-house computers saw an even more dramatic increase. It was almost an explosion of technology and dramatically changed all the methods that had been used until then. Part of the reason was cost. Many calculations take a considerable time even for a computer to work out, and if that time had to be paid for at realistic rates for an external computer, then the number of lengthy calculations was severely limited.

With their own facilities, the department was under much less restraint and complex models could be developed which could run all night without even printing a line of results until some time the following day.

As these facilities increased, so did their application. Foresters could now examine the future in a way that had not been possible before and today they are an essential tool in modern forestry. It is now possible, for example, to examine an entire district, know what is growing there, know when it will be harvested, know the sequence in which the harvesting will be carried out, and even know which areas are not possible to harvest after heavy rain.

Like mensuration and modelling, maps have always been a part of forestry. The obvious need was for people to know where they were or how to get to a certain part of the forest, bearing in mind that most forests are a considerable size. But there was another use. The generation that planted trees, certainly in the past, was not the generation that harvested them and maps were a means of recording the information of one generation for the benefit of the next.

The techniques used in making these maps changed very little until the mid 1950s. Until then, surveying gangs worked in the field using traditional techniques of chaining and triangulation and the maps were prepared from their field books. This was a laborious business. The plot was laid down on a plastic base called astrafoil and the detailed 'fair drawing' was made by hand on the matt side

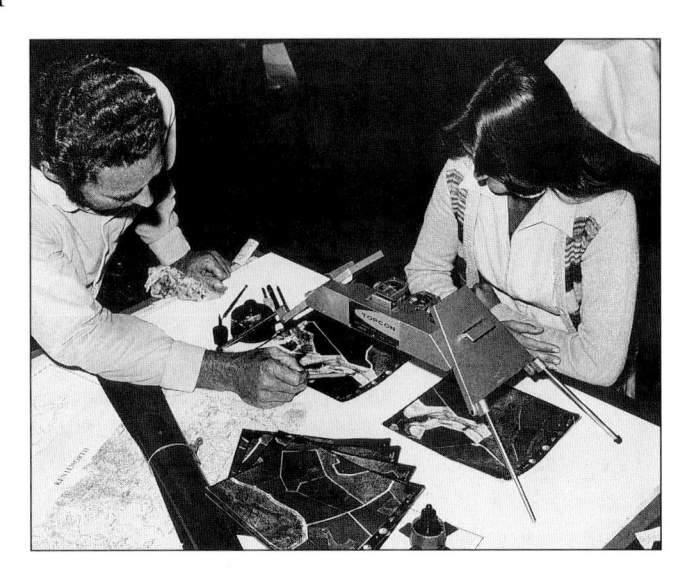

Experts using a viewer to examine a stereo pair of aerial photographs.

of a sheet of blue linen which had a coating of wax. All the lettering on the map was done by hand on the linen. Everything that eventually appeared on the map was drawn or written by hand using ordinary drafting instruments.

The first major advance came in the 1960s with the invention of photo-typesetting and dry transfer lettering. Both reduced the need for hand lettering as well as making the job much quicker. They also reduced the art and skill that had previously been required.

Even more dramatic, however, was the use of aerial photographs instead of surveying on the ground. This came into use in the 1940s, although an earlier forester, Jules Tardent, was the first to use it in Australian forestry: 'About 1932 I dashed up from leave in the south by rail to Townsville to board the aircraft carrier *Albatross* and carry out Australia's first aerial forestry survey from a pair of three-ton Sea Gull flying boats—maximum speed 80 knots! The navy's aerial photography installation at that time left a lot to be desired but the aerial mapping and visual timber estimating experience was of some value.'

To aid in the timely updating of maps, particularly of plantation areas, the department started its own programme of aerial photography in July 1977 using a 35 millimetre Canon camera mounted on a crude mount made out of pipes. In March 1981 this was replaced by a 70 millimetre Hasselblad carried on a specially designed mount.

Since 1977 most of the plantation forest estate in Queensland has been mapped from aerial photographs. Surveyors now spend much of their time working in conjunction with the aircrew doing the photography. Aerial photography is expensive but this is more than offset by the costs saved in reducing the number of surveyors needed in the field.

Aerial photography is still used for all forestry maps in Queensland. Satellite imagery, although useful for other purposes, is not suitable for making detailed large-scale maps as it still lacks the resolution needed for that purpose.

The next dramatic development in map making came in the 1980s with the use of computers. At first, simple drafting programs were used specifically to produce maps of plantations. These programs were followed by the much more sophisticated Geographic Information System (GIS). GIS technology uses digital techniques to store, display, manipulate, analyse and produce multiple layers of geographic information. Its introduction started a new era in map making.

Involvement in this system did not begin in earnest until early 1988, when it was recognised as being essential in order to cope with the vast quantity of geographical and biological data needed for a detailed assessment of the northern wet tropical rainforests.

The survey and mapping branch of the department developed this system using ESRI ARC/INFO software and using the flora and fauna data base that had been compiled by the Department of Environment and Heritage.

The initial area of data capture was nearly 900 000 hectares covering the proposed World Heritage area between Townsville and Cooktown. However, this was later extended to the boundary of the wet tropical rainforests biogeographic region, which increased the area to more than 2.5 million hectares.

This increase led to the installation of a Sun 3/60 workstation which could store much more data. It needed to, for at the height of the study there were about 3.5 gigabytes of data contained on thirty-six different layers. Additional data captured by space shuttle photography was also used to produce graphic 3D models for slope and aspect analysis of the more environmentally sensitive areas.

Although the development and use of GIS involved an appallingly steep learning curve for all those involved, its use has since been extended because of the advantages it offers.

A map can now be produced entirely on the computer. The base map of the required area is called from the data base held by the computer. The information that is to be shown on the map is also drawn from the data base and stored on separate overlays. In a simple case, for example, the base map for, say, the Atherton district will be retrieved from the data base. If the map is to show the location of

The draughtsman is using a digitiser to transfer map coordinates into a computer.

rainforests, then this information will also be retrieved from the data base and stored on its own layer. This will be followed by whatever further information might be required on the map, such as contours, rainfall and so on.

If the information required is not yet included in the data base, it can be captured by digitising information from another source. This is done by laying the source on a special table and then 'reading' it with a mobile device similar to a computer mouse. As well as being used in this map, the digitised information will be incorporated into the data base for future use as needed.

When the map is finished it is printed out either on a pen plotter or an electrostatic printer which is able to apply areas of solid colours. Maps printed by either method can be the finished product if only a small number are required, or they can be used as camera-ready art for large-quantity printing.

The system is both fast and flexible. The base map and the layers can be set to any scale required and a relatively simple map can be produced quite quickly. Simplicity in this case means the number of layers that need to be used, or whether one or more layers have to be digitised. What this means in practice is that a forester in the field can request a map of part of a plantation and it can be supplied in only a few days, or even less if the need is urgent.

Although the Department of Lands is the state mapping authority, it tends to concentrate on mapping property and this is why the forestry department produces its own maps. These maps are produced primarily for management, planning and fire protection purposes, although there is an increasing need for maps to support decision-making on environmental planning issues.

The GIS Services section now provides a wide range of information to management and those in the field. Different units cover native forests, plantations, technological development (which includes training others in the use of GIS), and cartographic services, which compiles and maintains all the official plans of State Forests and timber reserves as well as maps for management and fire protection purposes.

The maps that are produced are determined by the users, not the section, and there is a revision cycle of about seven years, although this varies with need.

The public also buys forestry maps because they provide information that is not usually available elsewhere. Many forestry maps are now marketed by the Department of Lands and all are available at forestry offices throughout Queensland. Indeed, the public use of forestry maps is growing all the time as people realise their value and accuracy in remote places.

Technology is not static. New techniques will be developed which will either make present methods simpler, or make available new and more sophisticated methods. In the immediate future, the developments in cartography are likely to provide those in the regions and in the field with direct access to computer information. So the ranger at a State Forest will be able to produce the map he needs instantly by calling it up from the central data base. Similarly, in mensuration, foresters measuring the trees in a sample plot will enter the data directly into the data base instead of completing forms.

There are many other aspects of technology in forestry as well as these two. Some are specific to forestry, others are common to the administration of all large organisations, private or public. The paperless society seems as far away as ever, but there is no doubt that technology, and especially the use of computers, has dramatically changed the way foresters manage the forests.

15 The Future

In 1993 the forestry department had a surplus of income over expenditure of nearly $7 million, and this amount was paid into consolidated revenue. It was the first time since the early days of the department that not only had it supported itself, it had contributed a very significant surplus.

This was not profit, however. The surplus came from the harvesting of plantations, but the capital cost of establishing and maintaining them still shows as a debt to the loan fund. But the fact that the department is showing a surplus means that this debt can now be reduced.

The department in Queensland is now the second largest plantation grower in Australia. These plantations occupy about 175 000 hectares and are worth an estimated $700 million. Even so, about 35 per cent of the state's sawn log requirements are imported from interstate and overseas, although this is expected to drop in the future.

The vision of the early heads of the department, especially Swain, who clearly saw the need for plantations and vigorously established the programme, is now more than vindicated. Had they not done so, Queensland would now supply very little of its own timber and there would be few sawmills and timber processors to provide employment and meet local needs.

The development of plantations will continue in the future, but progress is likely to come from better techniques rather than by increasing the number of plantations. Most exotic plantings in southern Queensland, for example, now use the F1 hybrid which was developed from Caribbean and slash pine. This hybrid grows quickly, is wind resistant and can tolerate poor soil.

It seems probable that this type of tree breeding will play an even more significant role in the future. If trees can be produced to close specifications they will more reliably meet the needs of modern mills with their sophisticated equipment. The concept of 'designer trees' is not very remote—trees with a maximum distance between branches, for example, which will provide longer and straighter lengths of timber and which will pass through the mill with a minimum of time lost in adjusting machinery.

Other technical advances are more difficult to forecast. The invention of the chainsaw changed forestry almost overnight, but nobody saw it coming until it was there. The major events are not usually predictable and there is little one can do to plan for them.

'Straight line' developments of existing technology are much more predictable but they are usually less dramatic. Some have been suggested in earlier chapters. Some developments are within the scope of current technology, others will come as the technology expands. About the only thing that is certain is that things will not stay as they are.

There will also be major changes in administration, and some of these are on hand now. In line with government policy, forestry will be 'corporatised', which means it will be run as a business rather than as a government department.

The story of forestry in Queensland is as much about changes in community attitudes as about changes in forest technology, and these changes are not without irony. The department came into being in order to conserve forests at a time when the community as a whole saw them as an obstruction.

Many people who came to Australia as settlers did so because of the opportunity to hold land. In Europe, where almost every square centimetre already belonged to somebody, such an aim was virtually unattainable. In Australia there seemed to be enough land for everybody provided they ignored the Aborigines, which on the whole they did. Such ambitions were not likely to be checked because the land was growing trees. Trees indicated fertile soil, and this meant it should be made 'productive'. At that time nobody thought trees were productive, and in any case there were so many of them that they would never be exhausted.

The irony is that foresters now find themselves in the opposite

The view from Wild Horse Mountain on the Bruce Highway near Beerwah in 1952 (above) contrasted with the same view in 1993 (below).

role. They have made the forests productive, but the community attitude has changed. From being the destroyer of forests the community now sees itself as their preserver, and in extreme cases accuses foresters of destroying them.

In Queensland the department has fought three major

THE FUTURE

engagements and lost them all. These were Swain's attempt to preserve forests in the tropical north, which cost him his job; the much later attempt to contest World Heritage listing of the same area; and the defence of logging on Fraser Island.

All three had some things in common. One was that the department was arguing a case against the weight of public opinion, and another was that it was not willing to offer any compromise.

On the whole the department underestimated the public arguments that were mounted against it. The public was not suggesting that the department was a poor forester, it was arguing for an alternative use of the land that excluded forestry entirely. The department defended its professional competence, and defended it well, but that was not the issue.

In all three cases a compromise would probably have been possible, but none was ever contemplated. The department argued for the status quo, to be allowed to continue its activities as before in its existing forests. It offered no alternative and in the end lost everything.

In each case, and certainly in the later two, the department was concerned about the domino affect. It thought that if it made any concessions the demands would increase—this area today, that tomorrow—and there would be no end to them. It believed that if it were to continue to perform its function of supplying its essential resource then it had to defend every hectare of its forests.

While the department lost all three cases, the effects were hardly disastrous, although in the case of the wet tropics they were certainly serious. The World Heritage decision locked up the entire rainforest and resulted in much hardship. Forestry activity is now restricted to other native forests and encouraging forestry activity by private landowners. The industry in the region has survived, but it continues to face problems of resource supply of non-rainforest timber.

There was nothing in the training and work of forestry to prepare it for such a significant shift in public opinion. If the department had been challenged it would certainly have acknowledged that it was accountable to the public for the use it made of the public's forests. In practice, however, this was not a matter of daily concern. Foresters were the experts, they were convinced they were acting in the long-term interests of the public (which indeed they were), and they simply got on with the job. It never occurred to them that the public might want them to leave some of the forests alone and, worse, they refused to believe it even when that became clear. Forests are for producing timber. What do you mean, you just want to look at them?

There are, however, some positive aspects of all this, and they are important for the future.

Although relationships between foresters and conservationists certainly became strained, they never broke down as they often did in other states. People continued to talk to each other and civility was maintained, at least in public. Both sides continued to maintain diplomatic relations and as a result the dialogue is continuing.

The department also recognised that, rightly or not, they had to pay more attention to what the public told them, and this is now a secure part of its policy. The department involves the public in its use of forest land, it liaises with conservationists about forestry technology and genuinely tries to accommodate varying views within its ongoing need to supply the public with timber.

The department is now far more open that it ever has been. Its role has changed dramatically, but it is very aware of that. Certainly there are older foresters who have difficulty coming to terms with this, but as we have seen elsewhere in this book, younger foresters have little difficulty combining conservationist attitudes with the discipline of forestry. Apart from the more extreme views on either side, they rarely were incompatible.

So although these public conflicts were traumatic, a great deal of good came from them. The department had to reconsider its role for the future, and some conservationists recognised that there was much more to forestry than they had first supposed.

The department has seen more change in the last twenty years than it did in the previous seventy. Rapid changes in technology were taking place at the same time that community attitudes were changing. The technological changes were easier to accommodate because most were directed at making forests more productive, but the learning curve in many cases was very steep.

This was not restricted to forestry, of course. Technological change put great demands on people everywhere. What differed was that some were affected earlier than others. Aeronautics, for example, started to change rapidly after the war as a result of the great advances that the war had brought about. Printing, on the other hand, barely changed at all until the 1970s. Until then the process was pretty much as it had been for centuries.

This was also true of forestry. Techniques were refined, certainly, but apart from regional differences and the use of machinery a forester working in 1900 would have seen little difference in the work being done in native forests after the Second World War. The same forester would see little that was familiar now.

So, what of the future?

THE FUTURE

Will wood be replaced by a synthetic material? Will hardwoods be grown in plantations? Will private forests play a role? Will helicopters pluck trees from steep hillsides and carry them to the nearby mill? Will trees be designed to grow so quickly that the whole state could be supplied from a forest little bigger than a municipal park? Nobody knows.

What is certain, however, is that the forestry department will cope with any or all of these possibilities. It has the expertise and the breadth of skills to lead development as well as to respond to it, and it has a broad view of forestry which can accommodate new ideas and different views:

> The overall purpose of the Queensland Department of Primary Industries Forest Service is the sustainable production of forest products and services within a balanced conservation program which includes the multiple use management of State Forest lands in accordance with the long-term best interests of the community. We believe that management decisions should be attuned to community attitudes and we recognise the need for public consultation in natural resource management. It is our intention to willingly fulfil our responsibilities as members of the communities in which we operate and we endeavour to manage our affairs, both individually and as a Forest Service, with integrity and excellence.

It is a statement that should last well into the future.

Appendix One

Heads of the Queensland Forestry Department

Sometimes called the conservator

L. G. Board, 1900–1905
P. MacMahon, 1905–1910
N. W. Jolly, 1911–1918
E. H. F. Swain, 1918–1932
V. A. Grenning, 1932–1964
A. R. Trist, 1964–1970
C. Haley, 1970–1974
W. Bryan, 1974–1981
J. A. J. Smart, 1981–1985
J. J. Kelly, 1985–1988
T. Ryan, 1988–1993
Norm Clough, 1993–

Appendix Two

Definitions of Some Forestry Terms Used in the Conservation Debate

Although many technical terms are used when discussing forests, especially in the context of conservation, confusion sometimes arises because some terms have come to mean different things to different people, or because they are applied more widely than when used by foresters.

One example is *rainforest*. What exactly is a rainforest? It is important to know as the term is probably used more than any other when forests and conservation are being discussed.

The term 'tropical rain forest' came into use at the start of the century to describe a certain type of evergreen forest in lowland areas of high rainfall within the tropics. Even then the term was not fully defined. It simply recognised that such forests usually contained many species in a small area and they were mostly trees varying in diameter and height and these supported many climbers and epiphytes (staghorns).

The difference between a temperate forest and a tropical rainforest was so obvious that a precise definition was not necessary. During the conservation debates that started in Australia in the 1960s, however, the definition of rainforest began to achieve more importance. Much of the debate focused on conserving rainforests and so it became important to know just what a rainforest was.

The closer one comes to a definition, the more elusive it becomes. Some rainforests clearly are rainforests and nothing else, in which case there is little room for dispute. But some rainforests will contain species that are not regarded as belonging to a rainforest. In other words, the rainforest contains some intruders. These intruders might be in the long-term process of taking over the rainforest, or the rainforest species might be overwhelming them. Is this, then, a rainforest? And how many intruders can there be before it stops being a rainforest? When does it become an open forest containing rainforest elements rather than a rainforest?

There are a number of definitions of rainforest and although they all agree in some respects, they tend to disagree slightly on the more marginal aspects. Some, such as the one contained in the Fitzgerald Report on Fraser Island, are so broad that they might cover many forests that are clearly not rainforest: 'A forest of broad-leaved, mainly evergreen trees found in continually moist climates in the tropics, subtropics and some parts of the temperate zone.'

The Queensland forestry department defines rainforest thus: 'A rainforest is a closed community (projective foliage cover of the tallest stratum exceeding 70 per cent) ranging in development from semi-evergreen vine thicket to complex mesophyll [leaves of moderate size] vine forest. Emergent eucalypts, brush box and satinay are infrequent or absent.'

That definition, based on that of Emeritus Professor R. L. Specht, is not a botanical definition but a structural one. The key element is the 'tallest stratum'. This is the tallest recognisable layer of the vegetation formation and trees that emerge above this tallest canopy layer (the emergents) are ignored.

The density of the tallest stratum is also important. In a closed community such as a rainforest foliage of the tallest stratum will cover at least 70 per cent of the total area. Forests and open forests have a density of between 30 and 70 per cent, and woodlands and open woodlands from 5 to 30 per cent.

If brush box and satinay form a recognisable layer *above* the layer composed of species normally associated with rainforests, then it will be defined as a forest or open forest and not a rainforest.

It is in the treatment of emergents that definitions differ. Some regard the emergents as being typical of the forest as a whole and therefore use them to define the entire forest. Others, such as the one used by the department, ignore the emergents and look instead at the tallest *layer* of vegetation. If it has a cover of more than 70 per cent, then it is a rainforest. If the cover is less than that it is something else.

Here are a few other terms often used in the debate, together with the definitions used by foresters.

APPENDIX TWO

Multiple use: The management of a forest area for a variety of uses, some of them simultaneously, such as timber production, water supply, wildlife preservation and recreation.

Old growth: This term was coined by the conservation movement in America. When used by foresters it means a forest of any kind which is dominated by mature and overmature trees and which shows no evidence of disturbance by man.

Regeneration: The renewal of a forest by natural or artificial means.

Regrowth: A forest of predominantly immature trees, usually of about the same age, which has regenerated after logging, clearing, fire or some other event that has removed part of the forest.

Rotation: The planned number of years between the formation or regeneration of a crop and its final felling. Several cutting cycles may take place within one rotation. In a plantation it is the time between the original planting and the final harvesting.

Selection logging: The removal of selected individual trees or small clumps within a forest.

Stand: An aggregation of trees sufficiently uniform in composition, age, spatial arrangement and condition as to be distinguishable from adjacent communities.

Sustained yield: The amount of produce of given specification that a forest can yield annually (or periodically) in perpetuity. Sustained yield implies continuous production by achieving an approximate balance between net growth and harvest.

Virgin forest: A forest which shows no evidence or has no recorded history of logging or clearing by human activity.

Yield scheduling: A technique for determining the sequence of logging and the appropriate allowable cut within a forest, using growth and harvesting models.

Appendix Three

Organisations Associated with Forestry

Australian Bureau of Agricultural and Resource Economics
GPO Box 1563
Canberra ACT 2601

Australian Conservation Foundation
10 Jane Street
West End QLD 4101

Australian Forest Growers
(Queensland Branch)
30 Mirbelia Street
Kenmore QLD 4069

Australian Heritage Commission
GPO Box 1567
Canberra ACT 2601

Australian National University
Department of Forestry
GPO Box 4
Canberra ACT 2601

Australian Soil Conservation Council
PO Box 858
Canberra ACT 2601

APPENDIX THREE

Australian Wood Panels Association Inc.
16 Woodgee Street
Currumbin QLD 4223

Country Sawmillers Association
Lacey Road
Aspley QLD 4034

CSIRO Division of Forestry
PO Box 4008
Queen Victoria Terrace
Parkes ACT 2600

Department of Environment & Heritage
PO Box 155
North Quay QLD 4002

Department of Primary Industries and Energy Agriculture and Forestry Group
GPO Box 858
Canberra ACT 2601

Department of Primary Industries Forest Service
Forestry House
160 Mary Street
Brisbane QLD 4001

Forestry Division
Department of Primary Industries
GPO Box 1604
Adelaide SA 5001

Greening Australia
GPO Box 9868
Brisbane QLD 4001

Greenpeace
93 Leichhardt Street
Spring Hill QLD 4000

Griffith University
Division of Australian Environmental Science
Nathan QLD 4111

Institute of Foresters of Australia (Queensland Division)
GPO Box 1697
Brisbane QLD 4001

GROWING UP

Men of the Trees
PO Box 283
Clayfield QLD 4011

National Association of Forest Industries Ltd
PO Box E89
Queen Victoria Terrace
Canberra ACT 2600

North Queensland Logging Association
PO Box 61
Ravenshoe QLD 4872

Plywood Association of Australia
PO Box 8
Newstead QLD 4006

Queensland Conservation Council
PO Box 12046
Elizabeth Street
Brisbane QLD 4002

Queensland Timber Board
PO Box 2014
Fortitude Valley QLD 4006

Rainforest Conservation Society
15 Colorado Avenue
Bardon QLD 4065

State Forests of New South Wales
Locked Bag 23
Pennant Hills NSW 2120

Tasmanian Forestry Commission
GPO Box 207B
Hobart TAS 7001

Timber Industry Training Council
PO Box 15
Newstead QLD 4006

Timber Research & Development Advisory Council
PO Box 15
Newstead QLD 4006

APPENDIX THREE

University of Melbourne
Faculty of Agriculture and Forestry
Parkville VIC 3052

University of Queensland
Faculty of Agricultural Science
QLD 4072

Wilderness Society Inc.
97 Albert Street
Brisbane QLD 4000

Wildlife Preservation Society of Queensland
160 Edward Street
Brisbane QLD 4000

Bibliography

Attiwill, P. M. and Leeper, G. W. *Forest Soils and Nutrient Cycles*, Melbourne: Melbourne University Press, 1987

Baird, D and J. 'The Unkindest Cut: Malpractice in our Forests', *Habitat*, vol. 10, no. 3, 1982

Beck, Jill 'The Historical Record: The Changing Vegetation of Australia', *Forest and Timber*, vol. 22, 1987

Beckett, R. *Axemen, Stand by your Logs*, Sydney: Lansdowne, 1983

Birtles, T. G. 'Trees to Burn: Settlement in the Atherton-Evelyn rainforest 1880–1890', *North Australia Research Bulletin*, no. 8, Darwin, 1982

Bolton, G. *Spoils and Spoilers: Australians make their environment 1788–1980*, Sydney: Allen & Unwin, 1981

Bootle, Keith R. *Wood in Australia*, Sydney: McGraw Hill, 1983

Burrows, Robyn *Dairies and Daydreams*, Brisbane: Boolarong, 1989

Carron, L. T. *A History of Forestry in Australia*, Canberra: Australian National University Press, 1985

Chisholm, A. H. 'Forest Revenue' (and other articles) *Daily Mail*, Brisbane, 1921

Dargavel, J., Goddard, J. and Caton, S. *Allocating Forest Resources in Queensland: Guide to Legislation, Regulation and Administrative Practice*, Canberra: Centre for Resource and Environmental Studies, n.d.

—'The Political Detection of an Australian Forestry Perspective', *Australian Forestry*, 43 (1), 1980

Dudley, Nigel *The Death of Trees*, London: Pluto Press, 1985

Fisher, D. E. *Environmental Law in Australia*, Brisbane: University of Queensland Press, 1980

Fitzgerald, R. *From the Dreaming to 1915: A History of Queensland*, Brisbane: University of Queensland Press, 1982

Frawley, K. J. *A History of Forest and Land Management in Queensland, with particular reference to the north Queensland rainforest*, thesis submitted to Australian National University, Canberra, 1983

—*Forest and Land Management in North-east Queensland 1859–1960*, Canberra: ANU, 1983

Gilpin, A. *Environmental Policy in Australia*, Brisbane: University of Queensland Press, 1980

Groves, R. H. (ed.) *Australian Vegetation*, Melbourne: Cambridge University Press, 1981

Gubby, A. C. *Campbellville and Cedar Days*, Brisbane: Department of Forestry, 1976

Hitzke, Daniela *Amongst the Tall Timbers*, Mt Mee: Mt Mee Centenary Committee, 1984

Hope, Geoff, and Kirkpatrick, Jamie *The Ecological History of Australian Forests*, paper presented to the Forest History Conference, Canberra, 1988

Hopkins, Russell *The Beerburrum Story*, Brisbane: Department of Forestry, 1987

Hunt, A. J. 'A View From the Outside: community expectations of Australia's forests and forestry', *Australian Forestry*, 46(4), 1983

Hyne, J. R. L. *Hyne-Sight: a history of a timber family in Queensland*, Maryborough, 1980

Jervis, James *Cedar and the Cedar Getters*, Sydney: Journal and Proceedings of the Royal Australian Historical Society, 25(2), 1940

Johnston, R. *The Call of the Land*, Brisbane: Jacaranda, 1982

Kerruish, C. M. and Rawlins, W. H. M. (eds) *The Young Eucalypt Report*, Canberra CSIRO, 1991

Lawrence, G., Vanclay, F. and Furze, B. (eds) *Agriculture, Environment and Society*, South Melbourne: Macmillan, 1992

Lines, William J. *Taming the Great South Land*, Sydney: Allen & Unwin, 1991

Luke, R. H. and McArthur, A. G. *Bushfires in Australia*, Canberra: Australian Government Publishing Service, 1978

McLean, P. *Arboriculture in Queensland*, Brisbane: Government Printer, 1889

McMahon, P. *The Merchantable Timbers of Queensland*, Brisbane: Government Printer, 1905

Marshall, A. J. (ed.) *The Great Extermination*, Melbourne: Heineman, 1966

Meston, A. 'Genesis or Red Cedar', *Campbell's Monthly Magazine*, January 1934

Meyer, Athol *The Foresters*, Hobart: Institute of Foresters of Australia, 1985

Morgan, R. K. 'Lahey's Canungra Tramway', *Light Railways*, no. 54, Summer 1975–76, Light Railway Research Society of Australia

Myers, Norman *The Primary Source: Tropical Forests and our Future*, New York: Norton, 1984

Pechey, Enid and Peter, and Milne, Hugh *A History of the CSR Wood Panels Gympie Mill*, Gympie, 1989

Perlin, John *A Forest Journey*, New York: Norton, 1989

Powell, J. M. *Environmental Management in Australia, 1788–1914*, Melbourne: Oxford University Press, 1976

Powell, J. M. and Williams, M. (eds) *Australian Space, Australian Time*, Melbourne: OUP, 1975

Queensland Forestry Department (variously called) *Annual Reports 1905–1992*, Brisbane: Government Printer

Report of the Commission of Inquiry into the Conservation, Management and use of Fraser Island and the Great Sandy Region, Brisbane: The Commission, 1991

Report of the Commission of Inquiry into Fraser Island, Canberra: Government Printer, 1977

Report of the Royal Commission on the Development of North Queensland (Land Settlement and Forestry), Brisbane: Government Printer, 1931

Report from the Select Committee on Forest Conservancy, Brisbane: Government Printer, 1875

Report of the Senate Standing Committee on Trade and Commerce: Australia's Forestry and Forests Products Industries, Canberra: AGPS, 1981

Rolls, Eric *A Million Wild Acres*, Melbourne: Penguin, 1981

Routley, R. and V. *The Fight for the Forests*, Canberra: ANU, 1973

Rule, A. *Forests of Australia*, Sydney: Angus & Robertson, 1967

Seddon G. and Davis, M. (eds) *Man and the Landscape in Australia*, Canberra: AGPS, 1976

Shepherd, K. R. 'Managing the Forested Environment', *Australian Forestry*, 42 (2), 1979
Sinclair, John *Fraser Island and Cooloola*, Sydney: Weldon, 1990
Smith, J. M. B. (ed.) *Australian Vegetation*, Melbourne: Cambridge University Press, 1982
Smith, J. M. B. *A History of Australasian Vegetation*, Sydney: McGraw Hill, 1982
Smith, L.W. *The Trees That Fell*, Ravenshoe: the author, 1991
Submission to the Commission of Inquiry into the Conservation, Management and Use of Fraser Island and the Great Sandy Region, Brisbane: Department of Primary Industries, 1990
Swain, E. H. F. *From Savagery to Silviculture*, Brisbane: Queensland Forest Service, 1925
—— *The Timbers and Forest Products of Queensland*, Brisbane: Government Printer, 1928
Tardent, Henry A. *Forestry in Queensland*, Brisbane: Government Intelligence and Tourist Bureau, c.1925
Thomas, A. H. 'Forestry in Queensland', *Telegraph*, Brisbane, 1931
—— 'Timber and Forestry Industry in Queensland', *Australian Timber Journal*, 4 (9), 1938
Tracey, J. G. *The Vegetation of the Humid Tropical Region of North Queensland*, Melbourne: CSIRO, 1982
Vader, John *Red Cedar*, Sydney: Reed, 1987
Wadhan, Samuel *Australian Farming 1788–1965*, Melbourne: Cheshire, 1967
Wallis, N. K. *Australian Timber Handbook*, Sydney: Angus & Robertson, 3rd ed., 1970
Watson, Ian *Fighting Over the Forests*, Sydney: Allen & Unwin, 1990
Webb, L. J. and Kikkawa, J. (eds) *Australian Tropical Rainforests*, Melbourne: CSIRO, 1990
Werren, Garry and Kershaw, Peter (eds) *The Rainforest Legacy*, 3 vols, Canberra: AGPS, 1992
White, Mary E. *The Greening of Gondwana*, Sydney: Reed, 1986
Williams, Fred *Written in Sand: A History of Fraser Island*, Brisbane: Jacaranda Press, 1982
Williamson, Nancy *An Historical Analysis of Forest Management in Queensland*, Griffith University, unpublished, 1985

Index

Italicised page numbers indicate pictorial references

Aborigines, 8–12, 15–16, 32
acacia, 8
Acclimatisation Society of Queensland, 48
Act to provide . . . Protection of State Forests . . . 1906, 53
Adelaide, 138
Adult Trainee Scheme, 140
aerial photography, *213*, 216–17
Allies Creek, 187–90
Amamoor, 184
America, early consumption of wood, 29–31
Archer, Charles, 42
Armitage, Edward, 163–4
Associate Diploma of Applied Science (Forestry), 140–1
Atherton, *52*, 56, 114, 122, 129, 130
Atherton, John Grainger, 68–9
Atherton Tableland, 14, 39
Australia, European settlement of, 32–4
Australian Forestry School, 138
Australian National University, 102; Department of Forestry, 138

back burning, 151–2
Balts, 80–1, 85
banksia, 6
barracks, 81
Barron Falls, 39
basal area, 212
Beerburrum, *80, 90, 134*
Beerwah, 64, 131, 133, 134
Benarkin, 57, 114, 130
Bernays, L.A., 48
blackbutt, 56, 113, 163, 165, 169
Board, L.G., 52–3, 226
Booloumba, *100*
Boondooma, 189
Bracefell, David, 161
Bribie, *127*
Brooloo, *38, 66*, 128, 129
Bryan, W., 92, 94, 101, 226
bullocks, *38*, 42, *57*, 165, 178

INDEX

Campbell, Winton Woodfield, 68
Canberra, 138
Cape Tribulation, 96
Cardwell, 103
Caribbean pine, 134–5
cartography, 215–19
casuarina, 8, 11
Casuarinaceae, 6
cedar, discovery of, 34–6; cutting, 36–42; in Queensland, 39–42; attempts to grow, 112
Central Station, 164
chainsaw, 178
charcoal, 20
Charles I, 24–5
Chinchilla, *177*
clear felling, 120, 137, 165
climatic changes, 6, 12
Clough, Norm, 226
coal, 23
computers, in sawmilling, 180–1, 189; in forest modelling, 213–15; in cartography, 217–19
conservationists, 94–9, 104, 119, 123–4, 168, 170–3, 224, 227–9
copper, 20, 22
Crete, 20–1
Crooke, John, 187–90
Crown Lands Alienation Act 1868, 44, 47
Crown Lands Amendment Act 1875, 50
Cryptotermes brevis, see West Indian drywood termite
cypress pine, 56, 113, 119, 144, 147
Cyprus, 20

Daintree, 39, 96
Dalby, 114
Dansie, Sam, 122–5
Dayboro, *57*
Department of Environment and Heritage, 217
Department of Primary Industries, 101

Douglas, John, 49
Durundur, 42

economic tree marking, 114
Edwardson, Captain, 159
Egleston, N., 30
emergents, 228
England, 21–9
enrichment planting, 116
Environment Protection . . . Act 1974, 167
environmental guidelines, 93, 94
eucalypts, 6, 11, 113
Eumundi, *49*
Eungella State Forest, 76
Evelyn, John, *22*

F1 hybrid, 221
Fedorniak, Victor, 85–8
felling, 40, *118*, 120, 137
Fellowship Certificate in Forestry, 140
fibre board, 176, 191, 199
fire, 142–56
fire boss, 152, 153
fire breaks, 144, 150–1
First Fleet, 32–4
Fitzgerald, Tony, QC, 99, 168
Fleetwood, Sir Miles, 24
Flinders, Matthew, 159
flitch, 180
Forest Boundaries Committee, 68
Forest Conservation, 1875 Committee, 39
Forest Learner Scheme, 139
Forest Measurer Scheme, 139
forest modelling, 208–15
Forest of Dean, 24
Forest Products Research Laboratory, 65, *195*
Forest Trainee Scheme, 139
forester, definition of, ix, 139
Forestry Act 1959, 83
Forestry Department established, 52

239

Forestry Technical Assistant Scheme, 140
Forestry Training Centre, 140, 141
Forests Products Bureau, 193
Forests Products showroom, 193
Fraser, Eliza, 159–61
Fraser Island, 56, 99, 112, 114, 128, 130, *151*, 157–69, 171, 179, 186, 223
fused needle disease, 133–4

Gallangowan, *143*
Gatton College, 141
Gilgamesh, Epic of, 18–19, 31
Gondwana, 5
Goodnight Scrub, 57, 114
Graham, John, 161
Greece, 20
Grenning, V.A., 73, 76, 77, 79, 80, 82, 91, 101, 226
Gresham, Sir Thomas, 23
Gympie, 61, 161, 182, 199; *see also* Forestry Training Centre

hakea, 6
Haley, C., 92, 226
Hawkesbury River, 35
Henderson, John, 40
Hill, Walter, 49
hoop pine, 56, 62, 75, 77, 78–9, *93*, 94, 111, *118*, 129–30, *136*, 147
Hyne, Chris, 186
Hyne, Henry James, 183–4
Hyne, Lambert, 184–6
Hyne, R. M., 51, 182–4
Hyne, Richard, 186
Hyne, Warren, 186
Hyne & Son, 163, 166, 182–6, 197

Illawarra, 36–7
Imbil, 57, *76*, 94, 112, 129, 185, *208*
improvement felling, 112–13
Injune, 62

Interstate Forestry Conference 1911, 138
iron smelting, 22–3

James I, 23–4
Jardine, John, 48
Jolly, N. W., 55–7, 58, 61, 62, 129, 137, 226

Kalpowar, *78*
Kamerunga, 112
kauri, *2*, 56, 112, 132
Kelly, John, 101, 226
Kenilworth, 86–8, *94*
Kessell, Stephen, 79
Keto, Aila, 170–3
kiln drying, *179*, 196–7, 205

land needed for settlement, 67–71
Lands Acts and Other Acts Amendment Act 1957, 83
Lands Department, 46, 47, 50, 51, 52, 67, 83, 218, 219
Laurasia, 5
loblolly pine, 133
logging, 177–8
logs, loading, 42, *160*, 178
Lynch Crater, 14

Macalister, A., 50
McDowall, District Surveyor, 51
McKenzie, H., 164
Macleay River, 37
MacMahon, Philip, 53–5, 72, 226
McNaught, Andy, 203–6
Macedonia, 20
mallee, 8
Manning River, 37
maple, 112, 194
maple silkwood, *115*
Maroochy, *181*
Maryborough, 134, 161–2, 163, 182–6, 201
Massie, Henry, 39
masts, England's need of, 25–9
Melbourne University, 141

240

INDEX

mensuration, 209
metrics, use of, ix
Mitchell, Thomas, 10, 12
Mossman, 39
Mount Gambier, 203–4
Mount Lindsay, *54*
Mount Windsor Tableland, *115*
multiple use, 92, 121, 144–5, 229
Myrtaceae, 6

Nanango, *93*
National Parks, 53, 55, *74*, 77, 92, 94, 167, 169
National Parks and Wildlife Service, 92
native forests, 109–21
Newcastle, 36
nurseries, 137

oak, 21–2, 110
old growth, 229
Otter, Lieutenant, 161

Palm Valley, *74*
Pangaea, 5
particle board, 199
Payne, William Labatt, 68
Peacock, Mark, 102–5
Pembroke, Earl of, 24
Perlin, John, 18, 28
Petrie, Andrew, 161
Petrie, Tom, 162–3
Petrie, Walter, 164
Pettigrew, William, 46–7, 162–3
pioneer species, 14
plantations, 126–37; first commercial, 62, 130; weed control, 73–5, 132, 137; site preparation, *90*, 135–6; use of exotics, 133–5; burning, 147; harvesting, 175–6; mapping, 216; future developments, 220–1
planting, techniques, 62–3, 135
plywood, *97*, 176, *192*, 194, 197–8
pollen, 14
pollen, fossilised, 7

Port Macquarie, 37
Port Stephens, 37
prescribed burning, 143, 144, 145–8
Primary Industries Corporation Act 1992, 101
Proteaceae, 6
Provisional Forestry Board, 63–4, 73

Queensland Conservation Council, 171
Queensland Forest Service, 101

Railways Department, 62
rainforest, 8, 94, *111*; in climatic change, 12–15; managing, 116–19; treatment, 117–18, 122–3; definition, 227–8
Rainforest Conservation Society, 171
regeneration, 229
regeneration felling, 113
regrowth, 229
Richmond River, 38
Rome, 20
rose gum, *54*
rotation, 229
Royal Commission 1931, 68–72
Rural Fires Act 1927, 144
Rural Technician (Forestry) Certificate, 140
Russell, Henry Stuart, 161
Ryan, Tom, 101, 226

Sample Tree Library, 209
satinay, *158*, 163, 165
Savage Report 1986, 99
saw pit, 40–1, *43*, 179
sawmills, 62, 64, 75, 77, 82, 164, 174–86, 187–90
Schlich, Sir William, 55
seasoning, 194–6
Second World War, 78–9, 201
seed orchards, 91, 137
seed storage, 75, 132

selective logging, 116, 120
shipbuilding, 21–2, *25*, 110
silviculture, 110–11, 113, 115, 126, 165; in rainforest, 117
site index, 211–12
skylining, 179
slash pine, 91, 133, *134*, 144, 147
Smart, Jim, 101, 226
snigging, 42, 75, 164, 178, *181*
snow gum, 8
Specht, Emeritus Professor R.L., 228
spotted gum, 189
springboards, *35*, 178
stand, 229
Standish, Arthur, 23
State Forests, 55, 57, 77, 79, 94, 165, 166, 219
State Government Computer Centre, 215
stumpage, 60, 66
Sub-Department of Forestry, 73
surveying, *211*, 215
sustained yield, 75–6, 119–20, 210, 229
Swain, E.H.F., 58–72, *59*, 101, 111, 113, 129–30, 131, 137, 193, 223, 226
Sylva, 22

tallowwood, 163
Tardent, Jules, 216
tea-tree, 6
tectonics, 5
Tench, Captain Watkin, 33
Theodore, Acting Premier, 67
timber regulations in Queensland, 39, 43–4, 57
Timber Reserves, 47, 94, 219

Toolara, 81; fire at, 153–6
Trade Department, 62
training, 138–41
tree breeding, 221
tree marking, 117
Trist, A.R., 91, 92, 226
Trist, C.J., 77
Tuan, 182, 186, 197
Tuan Creek, 134
twig borer, 112

Universal Wood Index, 193
University of Queensland, 141, 171, 215

veneer, 176, 197–8
Villeneuve, *71*
virgin forest, 229

walnut, 194
waratah, 6
Wayper, Jim, 199
Weatherhead tube, 62–3, 130
West Indian drywood termite, 200–2
Wet Tropic Management Authority, 103
Wild Horse Mountain, *222*
Wilson, Hart, 163, 166
World Heritage listing, 94, 96, 124, 169, 171, 223

Xerophytes, 6

Yarraman, 57
yield plots, 56
yield scheduling, 229
Yurol, *118*